THE RENOVATED HOME

THE RENOVATED HOME
Redesigning, Reorganizing, Redecoration

ANDREW WEAVING
with photography by Andrew Wood

HD
HARPER DESIGN
An Imprint of HarperCollins*Publishers*

THE RENOVATED HOME
Text copyright © Andrew Weaving 2005
Design and photographs © Jacqui Small 2005

First published in 2005 by:
Harper Design,
An Imprint of HarperCollins*Publishers*
10 East 53rd Street, New York, NY 10022
Tel: (212) 207-7000; Fax: (212) 207-7654
HarperDesign@harpercollins.com; www.harpercollins.com

Library of Congress Control Number: 2004114570

ISBN: 0-06-072355-6

Printed in China

First Printing, 2005

Half title page In Amsterdam, a full height pivoting door divides the areas while a flush fitted storage area is divided into sections shown by its doors.
Opposite title page This ex-industrial space has been divided diagonally from its neighboring unit. The steel staircase zigzags up to the new rooftop extension.
Contents page The massive beams of the previous train depot remind you where you are. Retaining these features gives scale and can divide up areas visually.

CONTENTS

THE ESSENCE OF
THE MODERN HOME

Page 6 A Victorian house full of original features. Here, tongue and groove walls, molded architrave, and trim and sash windows are updated with classic Scandinavian modern furniture. The steel-framed chairs are by Poul Kjaerholm.

1 Custom-made furniture by Bill Mostow for his mother's house gives a uniqueness to this bedroom. Furniture can instantly modernize your home. **2** In this 1960s house, classic pieces of furniture by masters such as Le Corbusier and Mies van der Rohe are placed within the rooms to divide the space.

In the words of Florida-based architect, Gene Leedy, "the urge to remodel is stronger than the sex drive." Having updated, extended, and renovated many of his earlier homes built since the 1950s, he knows what he is talking about, and in my experience, this is so true. Nearly every house I have worked on over the last twenty years or so has been bought by someone who loves it, loves every detail of it, loves the finishes, adores the storage, is eager to move in straight away, and begs me to leave certain items. But when you bump into a neighbor who tells you "they've stripped out that wall of storage," "they've painted all over your color scheme," or even that "they are putting in a new kitchen," you wonder why they paid so much for the house in the first place if they wanted to put their own mark on the place. This is what it all really boils down to: making your own mark. It makes the house yours; you can say, for example, "Oh, we've chosen these tiles for the bathroom, they were made specially for us in a little pottery in Tuscany."

Whether you are a do-it-yourself expert or not, this happens at all levels. I have even heard of cases where owners of brand-new apartments have changed fixtures, taken out walls, and laid new floors. You wonder why they bought the property in the first place!

Looking back, there have always been certain types who have searched high and low for a "wreck" of a house to do up. Now, with the fixer-uppers all snapped up, it seems that as long as the house is in the right location, even if

3 Graphic pillows, modern lighting and comfy seating all contribute to the modern home. In this city mews house the ceilings have been stripped bare to reveal the structural elements of the Victorian building.

4 Built-in furniture needs to be well planned and well thought out. Different levels give multiple uses, for example, organization and display, as achieved by the array of drawer fronts in this orderly dressing room.

it is in great condition, modernizing a house to make it work for you is today's lifestyle message. It is quite amazing how many people still go with what they initially see when they look to buy a new residence. I suppose they are buying into a lifestyle, even though when they finally get the keys, all they are buying are four walls with some fixtures and the leftovers of half a color scheme of the previous owners. To see past this is quite hard for some people.

Today we can do anything we want in our homes, within the guidelines of building codes regarding any planning issues and regulations regarding safety and construction. However, if you want to move any heating vents you can; if you want to make an open plan space, you can. The first thing to figure out is how this place is going to work for you personally and for your family.

When you see a realtor's ad with details saying "needs updating," "in need of modernization," or being sold "as is"—what does it mean? It all depends at what level you want approach the renovation. To some, "going modern" means that they are going to paint all the walls white, put down a light colored or natural carpet, or lay a wood laminate floor, but, within a warren of small rooms with bad lighting, next to no storage, and the wrong furniture, there is more to "going modern" than a quick coat of paint, even though that will help. Overall, there are different degrees of modernization. The simple use of color, finishes, lighting, and furniture can drastically change your interior. The reorganization of the space can revolutionize your home: a bigger living area, an open plan kitchen/dining area, or a wall of storage can make your home now work for you. Extending your space can give that extra room, floor, or level that you require. Whether you expand into the roof, excavate the cellar, or extend into the backyard, this degree of modernization can be most stressful, but most rewarding when completed.

Even on a very low budget, a few simple elements can produce amazing results. Many people do a lot of work before they move into their new home and yet end up changing many things once they move into the space when they realize that certain elements are not working for them. In my experience, it is crucial to live within the space first, experience the lighting changes, and work out a use for each room. Since the interiors magazine selection available now is ever increasing, with so many ideas and tips for this and that, collecting images of things that appeal to you is an important part of any new work you are considering doing on your home.

Modernizing your space depends on your needs, the type of accommodation involved, the size of the family living there, whether or not you work from home, the size of your budget, and how long you are planning to stay put. Are you doing the modernization for yourselves, or to make the place more desirable for a perspective buyer in the near future? Are you going to work on the home

5 An open-plan staircase of steel and laminated walnut rises to various levels. The glass partition acts as a safety barrier and allows light to flood through to the lower floor. **6** In this remodeled house in Brooklyn, New York, the lower level has been opened up, and access is via this staircase which also acts as important storage and a wet bar for the entertaining area. The stair treads were stained with Indian ink prior to sealing.

as one unit, or are you more concerned with the living, dining and kitchen areas? All these questions need to be considered in the initial planning stages, whatever level of modernization you desire.

Modernization also depends on what you have to start with. If the 1950s split-level house you have is full of original features, it would be a shame to remove these character elements. A mix of old and new works well if done sympathetically. An open-plan space with shuttered windows, fireplaces, and perhaps some paneling would be a great backdrop to some of today's modern furniture and lighting. Getting a balance of elements used is a very important factor. Whether you have an icon of modern design, such as a Mies van der Rohe Barcelona chair or a new design piece from Ikea, balance is what it is all about.

The scale of modernization is also to be noted. In a Victorian house, keeping many original features and perhaps replacing an internal wall with glass bricks may work, but the lighting would have to be considered, and the way the two periods unite would have to be well defined. More importantly, would this installation add to the overall design, layout, and use of the areas in question? Using certain "modernizations" merely for visual effect can make some spaces very aesthetically pleasing, but every addition really needs to have its own use. Of course, if the glass brick wall is going to bring much needed light to an interior hall, this is a good result, as this is both visually exciting and provides the solution to a dark space.

With the availability of so many products off the shelf these days—from places like Home Depot, Lowes, and

7 The long kitchen island is reminiscent of Pawson's work. Curtis Wood used Pawson's house as an inspiration for the client. The wall of doors opens to reveal more workspace and storage.

8 In this simply modernized home everything has been painted white—from the floors and walls to the vintage seating arrangement. The lightbox is left over from a recent exhibition. In this space the areas are divided with simple and effective sliding panels.

Target, doing-it-yourself is now much easier. Many stores offer printed instructions of how to do certain tasks, for example, how to replace a particular fixture, how to lay a wooden or tiled floor, how to put in a shower cubicle, or even how to build a glass brick wall. In some ways, this is a good move forward; but it could be potentially disastrous. To work on a series of individual areas is not really a good idea. On the whole your home is one unit and to make the modernization work, it is best to treat it as a whole. Even if you are simply changing the floor treatments—a consistent type of flooring on at least one level creates new modern feel to the place—it is the start of a total renovation. This includes all the other aspects to consider when modernizing an interior as a whole, for example, deciding on a color scheme for the entire space and using lighting that is similar throughout. If you work on a schedule and try to get things done in stages, the whole space will unite to be a modern place to live and work.

In some countries, the lack of modern housing has created an enormous interest in making what you have more modern. In a modern house it's usually the layout and room proportions that make the house more modern than a house of an earlier period. Reorganizing the space of your home can make a drastic improvement to any unit when it is done for a useful reason. There is nothing worse than an open-plan space that has had every wall removed and where the front door opens directly onto a massive void. Walls are as important as the space around them. In many cases, a wall in the right place can help create space visually. It is what is behind the wall that gives a feeling of a greater space beyond.

The most important part of changing the layout of your space is to understand the construction of what you have. Some walls can be removed without any structural supports, but others that are load-bearing need to be reinforced with beams. Advice from a structural engineer may be necessary. For example, in a three-story house with four bedrooms, there is the question of the space that has conventionally been used for sleeping. In an inner city environment, would a family with three or four children want to live in this location? Possibly not, so if you are working on this type of house, maybe more living areas could be created from some of the bedroom spaces. These days, even though some realtors still go on about the accommodation—4 beds, 2 bath—this is slowly changing; a 2-bedroom house with 2,000 square feet of living accommodation sounds much more appealing and spacious.

The division of space is what it is all about. In an "open-plan" living, dining and kitchen area, visual breaks are necessary to define the space, whether it is with different flooring to demarcate different areas, full height screens or partitions to break the view, or lighting elements to create pools of light where necessary. It is important to consider all these elements when stripping and remodeling your home.

To many, extending their home is the ultimate ambition. If you have a large unused roof space, a dark damp cellar, or a large side– or backyard, extending outward can save you from having to move. Of course, all these projects would need guidance from the appropriate authorities. Certain works are permitted developments, for example, extending upward to a certain size. Developing the roof space into living accommodation is all about access and ceiling heights, so if you have to raise the roof level, you would need a permit. With many firms now specializing in specific types of extension, the whole process is much easier. If you want to go for a glass extension, I recommend that you go to the specialists. They can do all that is necessary, from the plans to the finished results. These disruptive methods of making your house more modern and increasing your space really only work well when they are well thought out.

It has taken quite a long time to get to this urge for modernization. Not so long ago, it was the fashion that once you had become successful in your job and you had a chance to decorate your home, you would opt for a traditional interior of overstuffed chintz sofas and a few antiques—visual indicators that you had some money and good taste. Now there seems to be some kind of competition as to how modern you can make your home. You can look at "so called" modern interiors featured in magazines, but only once you have actually been in one and experienced the space do you realize that the same look might work for you.

The modern interior is really quite timeless. It is a matter of understanding your space, and making it fit your lifestyle. There are different levels of materials you can use to achieve the same desired effect. A painted concrete floor can give a similar feel to a honed stone floor, sanded and polished old floorboards can give warmth similar to an exotic wood floor. Putting together many elements that balance with each other create a modern harmony. Looking back to earlier examples of the modern interior, many places included the use of very basic materials, which were both easily available and just right for their use and effect. In the 1930s, architects such as Wells Coates and Serge Chermayeff were commissioned to remodel and modernize city apartments and houses, bringing them up to date with modern furnishings, new layouts, and lighting. With so much new building going on throughout the 20th century, the American modern interior has been part of a way of life. We look to these interiors for inspiration today, for color schemes and finishes. Mies van der Rohe gave us inspiration for placing furniture within the space, while Frank Lloyd Wright gave ideas for built-in furniture, recessed lighting, and the mix of materials used throughout the modern interior.

So whether you have a country cottage, a town house, a loft, or a brand new apartment you can make each of them better by understanding what you have and working out what you want.

9 A classic spiral staircase holds its own in its own unique space. The dark painted wall behind adds to its visual impact. **10** In this Florida home a wood burning fireplace has been installed to create a focal point. Even in the Sunshine State, a bit of extra heat is sometimes welcome.

THE HOME
TRANSFORMED

Page 18 Two neighboring apartments have been combined to form one unit, a rare occurrence since both apartments are owned by the same coop they were treated as one through the planning process.

1 A family heirloom—a patchwork quilt—has been made into a curtain. While making a great insulating drape, its visual impact unites the textures and colors used throughout this Helsinki apartment. **2** The Scandinavian 1950s screen-printed linen pillows and the wooden relief panel by Brian Willsher are the only elements of decoration in this grouping. The wall light is one of the original features from the house that has been retained and restored.

DECORATIVE MODERNIZATION

The easiest and quickest way to make your home more modern is by using simple elements such as paint, wallpaper, new floors, lighting, and key items of furniture. Even if you are renting your space, you can achieve amazing results with just a bit of thought, research, and time. By taking what you have, changing what is out of date, and adding selected modifications, the home you have can be easily updated. Without having to worry about structural alterations or any major upheaval of day-to-day living and working, decorative modernization is ideal for those seeking a quick, up-to-date home.

Most of the areas you will be working on can be altered very easily and quickly. The problem in many cases is making the decisions. Previously, the choice of modern elements was always quite limiting. Nowadays, wherever you go, whether it be the local hardware store or a home improvement superstores, the range of decorative materials is endless. From lighting to cabinet pulls, from wallpaper to the selection of white paint shades, the time it takes to decide can be the most difficult of all the procedures to create a modern home.

Where to start is another difficult task. You move into a new home with an out-of-date kitchen, colored bathroom fixtures, various floor coverings, bad lighting, and little or no storage. So where do you begin? With decorative modernization, as no major alterations are going to be done, it is not much of a worry. If you paint the bedroom bright red or paper it with 1970s wallpaper and

2

3 In the kitchen/dining area of Karim Rashid's New York loft, the use of optical laminate to face the kitchen cabinets makes these flat planes decorative. The grouping of pots by Ettore Sottsass for Memphis adds some color to this sociable area.

then you decide you hate it only after a few weeks, you can simply paint over it or cover it with something else. But if you can experience the space first, noting the light changes and work out the best layout for the room in question, you can try sample colors at different times of the day and get a more positive result.

Working on the decoration while you are still living out of packing boxes is a good idea since you can camp out in one room while you work on another. If you have been in the same place for a while, it is more difficult to get started. Start with a major clear-out of all unwanted or useless items that have accumulated over the years. With this out of the way, storage should be the first thing to sort out. In an older property this can often be a problem; however, under-stair storage can be organized and fitted with shelving, while alcoves can be furnished with display cubes and low cupboards.

Selecting items of furniture that have more than one purpose also makes this type of modernization more successful. Beds with drawers underneath offer very convenient storage space.

Whatever scale of modernization you are looking to achieve, lighting is one of the most important aspects. In many cases, the installation of a great modern lighting fixture can transform an otherwise dreary room. You can go for coordinated lighting throughout or just concentrate on the dining area with a reedition classic. Consider the practical requirements of each space in your home and how best to light these areas.

The placement of great-looking furniture provides the ultimate visual impact to a room. Whether selecting from chain stores or hunting out a vintage classic, the right piece in the right place will make the home more modern. Equally, fabrics are a great addition to a modern home as they can be used in a multitude of ways; from adding texture to a bare scheme to bringing tonal harmony to the different elements within the space. Patterned or plain, fabric is essential to today's modern interiors.

The following case studies show how others have added basic modern elements to what they have in order to achieve a livable and lasting modern space.

4 In Rashid's bathroom, the floor is the reverse of the laminate units. This totally enclosed area is divided from the main living space with a translucent Perspex panel.

5 In the child's room of a New York loft, Marimekko prints have been used, bringing color and pattern to an otherwise unadorned interior.

A TOTAL CONCEPT

Karim Rashid, a New York-based designer of furniture, cosmetic packaging, household goods, and interiors, is of British and Egyptian descent. Having designed hundreds of objects for some 400 companies over the past 20 years, Karim Rashid does know what the public wants. Whether it is a carpet, a lamp, a mirror, or a bottle of hand soap, he gets it right. Rashid has noticed that wherever you go in the world, you see the same stores, consume the same drinks, and can buy the same pair of shoes. He believes that we seek new experiences in the way we live and use our interiors.

In his own home, which has white floors and walls with wallpaper in selected areas, we see a total concept. he has designed the majority of the items in the entire space. If he moved to a new location, he could pick up everything and install it there, and his new home would have the same look and feel. The lack of features—except for the full height windows across the front elevation—make this space ideal for this installation.

The only structural elements to the space are the division for sleeping quarters at the far end, and the long rectangle taken out of the length of the main area. This rectangle, which is set aside for the bathroom and utility areas does not

reach the full ceiling height, leaving the entire visual volume unaltered. The wall of this area is broken into colored panels and doors leading to storage areas. The use of color to highlight this structure accentuates its depth.

Every surface of every object is a different material, from the new stainfree durable textiles to the vinyls and other plastics that are popular today. The lighting fixtures—either freestanding or wall mounted—are statements on their own.

On the whole, this space has been treated as a blank canvas, a space where an entire collection of furniture, furnishings, objects, and other items have been put together—essentially an eclectic mix of vibrant colors, textures, and surfaces, but where each piece relates to the next. This can be very extreme but certainly creates a lasting impact and fresh experience. As no real structural work is part of the visual scheme, the interior can be created very quickly.

3

1 Karim Rashid designed this DJ-mixing pod for Canadian company Pure Design. The wallpaper is also by Rashid. **2** A total-look installation in his own space showcases the range of products he has designed. From the rug to the seating, the 2 tier tables to the applique wall light, the only item not designed by Rashid is the Bang and Olufsen sound system. **3** In the white box loft space, the rectangular area, highlighted by the colored recess above, houses the bathroom and utility areas. Behind the yellow Plexiglas panel is the main bathroom. The dropped ceiling shows the position of the kitchen and dining area.

SYMPATHETIC RESTORATION

This house on the Essex coast of England was designed by Oliver Hill in 1935. Part of what became a large estate of international-style housing, this home is one of the very few that were actually completed, and is part of the largest group of individually designed modernist houses built in the UK.

Because this house was the first completed, good original photographs published at the time of construction have enabled a sympathetic restoration and interpretation of Hill's initial concept. His archive mentions all manner of design elements and decisions, ranging from paint colors to suitable planting for hedges. In one letter, Hill praised the work of the young architect/designer Alvar Aalto, whose furniture he suggested should be used in the house. Therefore furniture and accessories designed by Aalto are included alongside other pieces, also designed and made in the 1930s. It may sound extreme to limit one's selection to a period of design. But in a way, it is common sense, as the scale and design of this furniture was produced at the same time as the building, when modernism was seen as a way forward. How much more modern can you get than a simple white box with simple structured furniture without adornment?

When this house was discovered about five years ago, it had not been inhabited for some years. It had been "modernized" in the 1950s and then again in the 1970s. Both of these renovations removed original elements and added features of those decades. Every surface was covered with patterned wallpapers and carpets. Removing them revealed the original pine parquet floors and plain smooth walls that curve into the ceilings without coving. Doors to balconies and terraces had been bricked up, and new valances and tiled ledges adorned the windows. Now the house has all its original elements. All unnecessary decoration has been removed. Metal-framed doors have been installed, and ship-like railings on the balconies have been reinstated.

In the main living room, with its 15-foot quarter-radius window, furniture has been selected to create an ideal room for entertaining and relaxing. The floors have been stripped and polished. Simple graphic-design rugs and furnishing fabrics for pillows have been installed, and vintage lighting has either been restored or reinstated.

Overall, if you are working on a 20th-century modern-style home, assess the elements of the period and try to include them into a scheme for today, with vintage furniture and furnishings.

In this 1930s coastal property, the original metal-framed strip window curves around the house. The furniture is all period, but gives the room a contemporary feel. The clean lines of the architecture is complemented by the furnishings. Note how the walls curve into the ceiling.

COLOR & PATTERN UNITED

In this Manhattan apartment on the 13th floor of an old midtown industrial building, you find an installation of color and pattern put together by the owner Jo Shane who resides here with her husband and their two children. Shane, an artist, has a natural instinct and put the initial concept together. Having a great sense of color and being able to place items together in a harmonious way is a great gift.

Alongside a selection of family-inherited mid-century classic furniture, new pieces of design and craftsmanship from local artisans have been added. The use of pattern among the pure and simple lines of the George Nelson–, Charles Eames–, and Florence Knoll–designed furniture and the interior architecture by Jurgen Rheim of 1100 Architects, gives a balance of scale and impact while bringing the otherwise retro look up to date. In the main sitting area the long sofa by Knoll is alongside the surfboard table by Eames. Both pieces of furniture

1 The bed placed along the windowed wall gives ample space in the room for regular boys' activities. the color scheme brings a touch of nature to this high-level city location. 2 Classic mid-century furniture is grouped in the relaxing zone. With new rugs and pillows, this retro area is right up-to-date. The vintage lamp makes the Nelson coconut chair the ideal place for a quick read. An Eames surfboard table is an ideal surface for a collection of small, treasured objects.

3 The multi-tone green room for Shane's son has a built-in desk area for homework sessions. The wall-mounted storage units house both school books and treasured items and trophies. The slatted bench is by George Nelson.

are resting on a modern interpretation of an ethnic rug with the equally ethnic embroidered pillows scattered to the side. This area is reminiscent of the original Eames' house in Santa Monica. A collection of different cultures and styles alongside modern design elements creates a pleasing effect.

The low-level approach is also reflected in the long low bench that runs along the entire length of the apartment. The George Nelson coconut chair, a great chair to sit in and to look at, brings together this conversation and relaxing zone.

The children's rooms are away from the main core of the apartment. The unusual shape of these rooms enables them to serve as both a sleeping zone and a place to work. The

bed along the window wall gives ample space to play and also provides seating when friends visit. The desk provides a place to do homework and ideally will help instill organizational skills from an early age. Storage and display areas are important for children as it means they can live, work, and play in an organized environment and make sure that they keep their own rooms neat and have their favorite things on display.

The various shades of green in this room bring the color of nature into the son's room. At this high level the apartment doesn't see any treetops. This room also includes useful modern design classics such as the Jacobsen series 7 chair, the Nelson slatted bench, and steel-cased nightstand.

CLASSIC INTERPRETATION

This loft space in London's docklands near Canary Wharf has been filled with mid-century modern furniture classics. Using furniture as a decorative way to modernize this home was the easiest way to achieve the best result. If only a coat of paint is required, then furnishings can effect an almost instant transformation. There are endless outlets for good and interesting pieces, whether by designers or not. In this urban space, the majority of items are American, including works by George Nelson and Florence Knoll.

As well as looking good, the furniture you select has to work, as space is usually valuable. In this space, a storage system by George Nelson has been installed along the wall of the main living space. This system, called CSS, Comprehensive Storage System, was designed in the late 1950s and developed over a few decades with update finishes and hardware. This particular system was produced in the mid-1960s. It includes drawers, display cabinets, drop-front desk units, book storage, and display shelves. It groups everything together using the enclosed areas for organization and using the open areas for display. On this unit a large selection of wooden sculptures by London sculptor Brian Willsher provide the key focal point.

When this loft space was purchased, the decoration was somewhat personalized by the previous owners. The walls and concrete floors were brick red. Now, with the white walls and the painted concrete floor, the space is now a simple environment for good-design furniture to be seen and used. The seating throughout the living area is all by Florence Knoll. The classic sofa and occasional chairs still retain their original upholstery.

In this simple makeover, pattern is bought in with graphic pillows by New Yorker Jonathan Adler and an old striped rug from a local flea market. A bit of pattern brings scale to the space.

If you are lucky enough to be able to go into a new space that just needs decoration, keep in mind how you want the place to look. It takes more than paint and textiles to make the place yours. You need to add those items you have found that you really like, to which you add good-looking, comfortable furniture and the right kind of lighting to create a sympathetic mood for the interior and highlight the treasures of your home.

This apartment shows how you can take what you have, place it in a new space and add new elements to create an exciting modern environment.

2

1 The white-painted brick walls act as a great backdrop to the vintage furniture and the skyscraper sculpture. **2** In the main living area off this loft space, a storage unit by George Nelson dominates the area. Here all is hidden away or on display as necessary. The concrete floor is simply painted.

This small apartment in Helsinki is on the upper floors of a turn-of-the-century block, typical of many European cities. Comprising of four-story buildings with two apartments on each floor, the block is set around a communal courtyard used for bicycle storage and hanging wash out to dry. In this space, designed by interior architect Ulla Koskinen, simple, complementary decoration has been used to add to the impact of the classic twentieth-century furniture which just proves that "modern" doesn't have to be brand new.

The apartment consists of two bedrooms, a study, the main living room and a kitchen/dining area. Typical of Scandinavian design, the floors are stripped wood with an opaque white wash to keep them pale.

In the predominantly white interior, primary colors have been introduced via furniture upholstery and lacquer finishes. The main sitting area boasts a long, low, red sofa running under the two windows and a pair of 1930s Aalto cantilever armchairs, bringing the modern elements to this period apartment.

1 A long, low sofa runs the length of the window wall. A deep shelf has been built in to increase the display depth, and unites the two windows while also disguising the heating elements for the apartment. 2 An Alvar Aalto chair brings some color into the white space. Behind it is the original ceramic fireplace. To the left are the double doors to the dining area.

3 Through the double-paneled doors is the kitchen/dining area. The red lacquer table brightens up the monochromatic scheme and makes the room less kitchen-like. **4** A section of cabinet has been removed to house the television for dinner-time viewing. Through the narrow door at the end is the bathroom.

This color scheme shows how, with the addition of color to your home, you can bring it up-to-date, while at the same time making the most of its classical proportions.

This space, with its original features, paneled doors, base boards, ceiling moldings, and a ceramic corner fireplace, remains as it was built. As all these details are painted white, they remain subtle and yet are key elements and provide a solid background for the new elements that have been introduced.

In the kitchen/dining area, a mix of black and white units makes the space more of an extension of the main living area than a kitchen, as the cabinets appear to be like built furniture. The bright red lacquered table was designed by Koskinen for this space and again brings fresh color to a monochrome scheme.

When trying to make a home more modern, people often overlook making the most of what you have. There is no need to remove period details. Painting everything one color disguises these features to some extent, but still reminds you that you are in an older surrounding. It is more about what you add to the space, than what you take away that makes the space up-to-date. For example, if the paneled doors had been replaced with flat ones or the ceilings had been lowered for low voltage downlighters, this space could easily have become just another bland "modern" apartment, void of any soul or character.

UNIFORMITY OF DESIGN

1 This main living area with its painted floor and furniture is white from top to toe. The red shag rug brings in color. Note the exercise equipment. 2 Stainless steel-fronted kitchen cabinets and appliances work well in this industrial space. The recess used is just the right size for all the necessities. 3 The dining area is divided from the sitting area by the original half glazed wall. The deep beams run through both spaces, as do the vintage enameled hanging lights.

This industrial unit in the Kruununhaka area of central Helsinki was once a knitting factory. Now taken over by creative director Mikko Mannisto, his fashion designer partner Tuula Poyhonen, and children Mosse, Olb, and Ere, it is a simple home for a family with one full-time child and the occasional visit from the others.

This space is all about going back to basics—what you see is what you get. The fairly small space of four rooms with a bathroom is on the ground floor of the 1920s development, originally built for businesses in the town center. Surrounding internal communal gardens, the external space is like a walk back in time. However, once inside it is a different story.

The entry is a large space divided from the next space by an original half-glazed screen wall. The compact kitchen area with its stainless steel–faced cabinets creates an industrial feel to complement the utilitarian space. This small grouping of elements contains all that is needed, from stove to dishwasher. The furniture appears to be discarded office chairs scattered around an old pine table on wheels. The simplicity of all the elements makes it easy to feel at home. There are no precious furnishings to damage and no "designer" chairs to sit on. It is all child-friendly and

complimentary to the building's original architectural and decorative features. Floors and walls are painted white, creating a simple backdrop where everything looks good.

The glazed double doors lead through to the main living area with a vintage sectional seating arrangement, painted white with white upholstery. Off to the side lies the simple sleeping area—futon and a storage facility on wheels—that is reached via a sliding partition. The only color found in the entire space is introduced with the use of throw rugs—a rag rug in the kitchen and a red shag rug for the living area.

Between the deep beams in these rooms hang original enameled steel pendant lamps. Using the same fixtures throughout unites the spaces; a uniformity of design and period is always important. Retaining so much originality gives it soul and a firm footing. On a modern note, the dividing partition adds a great feature; it divides the space, and the windows on both sides add to the ease of organization here with both sides well lit.

Overall, this type of space works when you consider what there is to start with. Using decoration to make your space more modern is nearly an almost instant transformation, whatever the original style or type of accommodation.

UPDATING FEATURES

In any type of property, some original features will exist, some will have been completely removed. The last house I lived in had been updated in the 1970s. All the features except the dog-leg staircase had been removed. So we put in a new modern staircase. The new flight of stairs created more space in the hallway and instantly united the modern kitchen area with the open plan living area.

Situations like these are quite hard to come by now. With older, more intact houses being in such demand in recent years, even if an older home was missing some original details, they most likely have been restored by now. In some cases you could be taking on a project that could involve a combination of removing and updating reproduction features.

The best way to start is to take note of what features exist, what features you want to keep or restore, and what features need to be updated. It is important to get a good balance of original and new features.

Keeping or reinstating the elevation of the house always works well because it gives the house its architectural balance of window space and masonry. Inside a room these windows give the correct balance of light and proportion to the space. If you intend to keep a few features, it is always best to retain the most original features, even if they are not completely intact or in pristine condition. Great molding and baseboard act a backdrop to a large abstract canvas.

3 A molded ceiling is updated with the use of a contemporary glass suspended light. While making use of the power supply, this type of light will bring much-needed illumination to the main living room where it is installed.

Removing moldings can be a lengthy and expensive job. Sometimes you may get half of a ceiling in one room and the other half next door. Taking them down will, of course, give a new feel to the room and remove the obvious signs of an earlier alteration.

The focal point of the main room will often be a fireplace. A simple marble surround can be updated with a simple grate. However, removing the fireplace completely can sometimes make the room appear quite bland. Keeping something there as a focal point is a good solution, and even a hole in the wall can look effective.

Sometimes changing simple things, like the height of doors to rooms, can dramatically modernize the space. Changing conventional doors to sliding panels can be even more radical and provide much more room. Equally, the unification of floor covering can give a great feeling of space and modernity. If you are lucky enough to have a great original floor, make the most of it and restore and repolish it. In many cases, floors are patched up and cut up. The easiest way to improve this situation is to install a new surface over the existing level with a new or reclaimed material that follows through from room to room. With so many choices available, ranging from massive 10" oak planks to poured resin, new flooring can be the quickest way to update your space as it unites the spaces and gives a simple background to whatever you add to make this space your modern home.

Sometimes features need to be removed when they dominate the scheme you are working on. This is especially true in reclaimed industrial spaces. Some structural features give great character to a place, but if you want a simple up-to-date space, a series of structural beams can be unsightly. Lowering ceilings to disguise these elements can also be a way to discreetly house lighting and sound systems.

Overall, updating the features of your home can be executed in many different ways. One option is to keep everything in its place but changing its appearance and finish. Repositioning the focal elements can not only give great effect, but also make the layout appear to have been altered, giving it fresh visual appeal. The case studies in this chapter cover many aspects of updating, ranging in scale and budgets.

4 Retaining the original features such as the newel post and turned banisters reminds you that you are in a Victorian townhouse. Under the stairs where the old dark bathroom lay, the compact space has been transformed into a bright glazed shower room. **5** This room within the center of the house has had one of its doorways filled with glass blocks. While bringing a modern element, it also allows an expanse of light into the living area.

PLAIN AND SIMPLE

1 In this open plan kitchen, a folding ladder provides access to the much-needed outdoor space of this urban mews house. When not in use, it can be pushed up into its recess. **2** Some of the exposed elements of the roof structure are visible above this kitchen space. The stools and high table create a handy place to perch for breakfast. **3** The kitchen is installed along the back wall of the house. Open to the rest of the space, it needs to fit in visually with the surrounding features. Using wooden doors on the cabinets ties in with the flooring and stairway details found on this level. The dining set includes a set of three-legged chairs by Hans Wegner.

A typical Central London mews house looks small from the exterior, like a two-story cottage. What must have originally been the garages are now the first-floor bedrooms, creating an upside-down house.

Upon entering the main living and relaxing area on the upper level, it is evident that the house has a wing at the back that doubles its perceived volume. Originally it would have probably been a warren of smaller rooms. Now it is treated as one space. The L-shaped floor plan divides the space naturally so the kitchen area spreads across the back wall, so that it is not visible, when relaxing in the entertaining area.

The ceilings and some of the walls have been stripped back to their basic construction materials. The house has been updated by creating a loftlike space. With the bare rafters and beams and the exposed brickwork, the house feels like an industrial or agricultural building. In contrast, new plaster has been applied in the kitchen, to give a slick and contemporary feel.

There is a good balance of old and new in this house. Only the dividing wall of the neighboring house has been stripped of its plaster. This wall includes one of the two fireplaces on this floor. The fireplace in the kitchen/dining area has a small period wood-burning stove. In the main relaxing area, a contemporary firehood stove has been

4 Within the roof structure, skylights have been installed to bring more light into the center of the open-plan space. They are screened here with a blind system on pulleys to diffuse the overhead midday sun. **5** The staircase comes up directly from the lower level into the main living space, alongside the brick wall that continues down the stairs. There are various skylights throughout the roof, bringing natural light to the whole space. The stair surround becomes a natural display area.

installed, replacing the original fireplace. This really updates this area, and, with the shag-pile rug, gives the impression of a movie set from the 1970s. Skylights have been installed in strategic places in the roof to bring light into the central areas of the reception floor. Blinds on the skylights cut out the overhead sunlight.

Throughout the house there is a mixture of vintage and contemporary-designed furniture and accessories. In the kitchen, a Hans Wegner dining set sits alongside a pair of vintage fireside chairs. Using a mix of new and old can give your home an instant 'lived in' look. Vintage pieces give soul to a new interior and bring individuality to a

modern environment. With so many of today's interiors resembling each other, it is prudent to spend time looking for more unusual pieces of furniture to achieve a unique look for the home.

This house has been updated simply by removing the decorative finishes applied to the construction materials. Revealing brickwork and beams gives a certain look and you need to be careful to keep from giving the space too much of a rustic feel. Featuring just one wall of brickwork shows the structure of the house without dominating the overall contemporary feel of the space.

Anne Fougeron, a San Francisco—based architect, lived in this house for many years before deciding to renovate it. She loved the Victorian house with its handcrafted features. So she decided to work with what she had and make the most of the proportions and details. Throughout the house, good simple modern design elements are successfully combined with the original aesthetics of the old house.

Fougeron didn't want to turn this house into another contemporary makeover. Taking note of what the house had to offer, she recognized that the areas that needed the most help were the dark hall, the bathroom, and the featureless back of the house. By retaining all the modernizations to the original footprint of the house, Fougeron was able to keep all the original services in their existing positions, even though the electrics and plumbing were updated at this time.

The original kitchen always lacked the inviting appeal that you would expect from a family kitchen in this kind of house. Removing the back wall and replacing the gridwork of the old windows with laminated sheets of glass opened the kitchen up completely to the outside terrace. While instantly increasing the size of the room, visually it also brought the

BEFORE LEFT
Beyond the main living rooms is a selection of smaller divided rooms, including a small bathroom with a separate toilet area. The kitchen includes a walk-in storage area.

AFTER BELOW LEFT
A new shower room has been installed and juts out from the existing flank wall, thus giving a greater floor area to the kitchen space. The solid walls between these rooms have been replaced with glass.

2

1 The kitchen, still in its original location, has had its back wall completely removed, opening up this entire space to the outside via a totally glazed wall. **2** The central kitchen island houses the cooktop. This is the hub of the household, with the barstools making it an ideal place to sit and socialize. On the left is the glass wall of the shower room.

3 & 4 The fully glazed shower
room has a translucent glass wall
that looks onto the kitchen. The
recessed strip light glows green at
night. 5 The well-fitted shower
room has a built-in stainless steel
vanity unit, recessed storage, and
ample mirrors. The terrazzo floor
butts up to the existing oak strip
floor which has been bleached
and polished.

outside in and opened up the back of the house to provide
great views of the neighborhood beyond. A minimal steel
frame with the glass mounted with clear silicon allows the
plane of glass to be almost uninterrupted. The floor-standing
units are complimented by the central cooking and eating
island, making this area the core of the house where
everyone can gather.

The hallway bathroom lies alongside the kitchen.
Previously a narrow, dark room, this space is now filled with
light, achieved by installing laminated glass in framed panels,
which mimic the full-height windows in the kitchen. The new
shower room is accessed from beneath the ornate Victorian
staircase. This bathroom is treated as a complete wet room
with its terrazzo floor and gully for drainage. There is also a
spacious built-in sink area and concealed steel-faced storage
cabinets. At night the opaque glass is illuminated by a subtle
neon light.

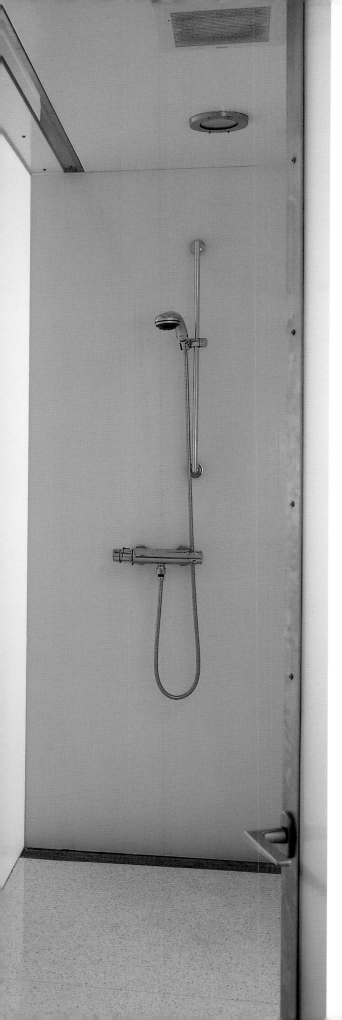

The work took about 12 months to complete, and in this time Fougeron was able to try new products and include well-tested techniques from her previous projects.

From the front of the property, it is hard to tell what size the house is or what its layout is. This house is very deep. The second floor consists of a large living room and a dining hallway that leads onto the kitchen at the back of the house where the terrace has access by a spiral staircase to the garden below. Classic pieces of modern design work well within the proportions of this historic house are found.

5

In the relaxing zone, the spiral staircase to the upper floors has been retained, but nothing else. The fireplace, which extended much farther into the room with brick piers and a copper canopy, has been updated with a strikingly plastered and painted chimney breast and storage for logs. Throughout the space are iconic pieces of design like the Arco lamp, now being reissued, that arches across the space.

The door at the end of this room leads to the original entrance foyer—partially divided with a wall of glass bricks that provide valuable light to the space—and the original heavy front door. Through the entrance foyer is the kitchen, the focal point of which is the original extractor hood, giving the cooking area a permanence. At the end of the kitchen, full-height windows open out to the courtyard. These doors and windows were originally painted black. They have been spray-painted white, so now they look as modern as the whole kitchen, and bring uniformity and brightness to the overall space.

This old schoolhouse has seen many transformations in its lifetime. Originally a Victorian village school, it has since been a pub, a tea room, and various other establishments in between. The layout of the house has been added to and extended over the years and consists of the original school hall with the principal's office off that and a one-story extension to the side and back. It has been saved by its new owners, Angi Lincoln and John Orum, who have tackled the lengthy job of clearing out the old and bringing in the new.

The ground floor now consists of a large living room with a spiral staircase to give access to the floor above, a central kitchen that unites the areas, a study room, and various other rooms for activities such as working out or painting. Three bedrooms, a dressing room, and a large bathroom occupy the upper level.

Many features that the owners inherited have been updated. The floor finishes throughout have been changed and upgraded, from the tongue-and-groove plywood panels in the kitchen to the laminated strip in the relaxing zone. Previously there were a whole range of materials. Keeping to a more uniform color and material helps unify the spaces.

3

1 The fully working fireplace acts as a focal point, as well as summer storage for logs when not in use. The cowhide rug provides pattern and texture, and the Arco lamp brings the schoolhouse up to date, even though it was designed in the 1960s. **2** Using the industrial-looking extractor ductwork from the old kitchen gives the new cooking area a professional look. Light floods in from the courtyard at the end of this space. **3** The existing entrance to the house still retains its old oak front door. This area has been modified to include a cloakroom divided from the original space by a glass brick wall.

COOL AND FRESH SPACES

1 The old kitchen cabinets have been faced with white laminate to update them. The wooden walls and ceiling have been painted white, and the floor has been refinished with bricks. **2** In the large master bedroom, a large shower suite makes the space feel complete. The full-height sliding doors give access to each side of the house as well as adding a natural cooling effect.

In this 1950s post-and-beam house in Florida, the entire space was initially updated and modified by the original architect, Gene Leedy, and since by its new owner Robert Kaiser. From its original plan it has been slightly extended, so that what were once two bedrooms are now one.

This home was originally quite obviously a wood-framed house, as early photographs show the walls to be polished veneered plywood, the ceilings were tongue-and-groove cedar, and the floors used to comprise of cork squares. Fifty years from its construction date, the house is now as modern as it was when first built. The almost transparent aspect between the interior space and the outer area surrounding the property is apparent, due to a series of sliding glass doors. Visually, they create a seamless blend of architectural features, for example, directly outside the bedroom is a new swimming pool, installed by Kaiser. Across from the pool a guest house provides relaxed accommodation for friends and family.

In the main house, the original cork floor has been replaced with a brick one. Laid in an alternate bond pattern, it relates to the concrete block used elsewhere around the house. This flooring runs the entire length of the house, thus treating the space as a whole, and offering color, texture, and a cool surface. Likewise, within the whole space, the walls and ceilings have been painted white, brightening the house.

The kitchen has also been updated. Once a wood-faced kitchen, it has now been paneled with white laminate. The bank of original wall cabinets on the back wall have been replaced with modern horizontal steel-framed cabinets. Also in this area, the original louvered, jalousie windows have been replaced with sliders. These offer ideal access for serving drinks directly outside.

In the bedroom area of the house, originally there were two double bedrooms divided by a wall of storage. With this partition removed, the space becomes one. Perhaps this is luxurious use of space, but with the guest house across the way, there is accommodation for visitors. One of the two bedrooms had its own bathroom; now the large master bedroom uses it. Full-height closets run the length of the new bedroom space. The sliding doors at both ends allow the space to be exposed to the sun both morning and evening, and, when the doors are open, the cross breezes help cool the room during hot spells. This house shows that an older space that was so modern when first built can easily be updated to be even more modern than originally intended.

REORGANIZING YOUR SPACE

Space is important, but its organization is even more so. Taking the overall space of your home, removing all existing walls and partitions, and reconfiguring the area as a whole is the way to get everything you want in the correct place for today's way of living.

If you have a blank canvas to work with, reorganizing the space can be quite a task. Where do you start? How open plan do you want it to be? What do you need to include in the layout? What is your budget? What is the look you are going for? What are your priorities? How many sleeping spaces do you need? How big do you want the kitchen to be? I have always found living in the space first is always a good start. But while you are experiencing the original layout, you also need to have some idea of how you perceive the space changing for the better. Good basic construction and knowing what you want are valuable factors in establishing what your options are, so your first consideration should be to call in structural engineers, and discuss with the local building office what you can and cannot do.

When reorganizing the space you need a list of requirements. How much storage do you want? What about focal points, display, and specific areas for furniture and artwork that you already own? If you are creating a completely new layout, you have no excuse for bad use of the space; to discover after a reorganization that you have nowhere to put anything; to leave the existing poor lighting where it is; or to not plan enough storage.

In some projects I have worked on, some of my clients' ideas have not been thoroughly thought out. Sometimes the idea of creating your own space by changing the layout gives people the idea of their home being completely open plan. I've always found that this never really works. With a total open plan space, there is never anywhere to put anything, every aspect of the space is in constant view, and on a practical level, it can be hard to organize the services. However, the installation of a partition in the correct place can make an area feel twice the size, and one is immediately curious to know what lies behind the wall. The division of a space with a partition of double-sided storage always seems to work well when you are figuring out the space layout and storage options.

To some extent, the end result is under some constraints from the original format. If you are working on an old row house, you can remove walls, relocate the staircase, and change the layout. The idea of moving the bedroom areas to the first floor and the living room upstairs works well for some people, as it might give a more spacious living area, particularly if you open up the roof space. It all depends on your lifestyle. With any house that is built in a row of houses, the width of the house does limit the new layout. Windows that are positioned only at the front and back of a property creates a railroad effect. This results in some rooms without windows being sandwiched between other spaces. Reorganizing the space can compensate for this, but the building department may see this differently. It would all depend what the rooms in question are to be used for, and there would need to be adequate light and good ventilation.

In an apartment, it is sometimes much easier to reorganize the space. A new-build unit would generally have many windows, and therefore you can plan the reorganization so that each designated area has access to a window and services where necessary.

The industrial space that may comprise of a shell loft apartment gives you a clean slate, and poses its own challenges. Personally, I find it easier to remove and relocate elements than to put up a wall in a completely empty space. In some ways, the ideal end result would be a space with a lot of flexibility, in which the dynamic changed

2

according to the combination of movable partitions, sliding walls, and versatile furniture within.

With many of us working from home now, to have a small area that can be a home office during the day, and extra sleeping space when needed, is an increasingly popular option. Including this type of work space is always an added bonus to any home. In a regular home layout, a total room allocated to a working space would be wasted space. So configuring how best to include this type of office should be one of your key priorities.

The spaces in the following case studies include houses, remodels, and industrial spaces.

HISTORIC MEETS MODERN

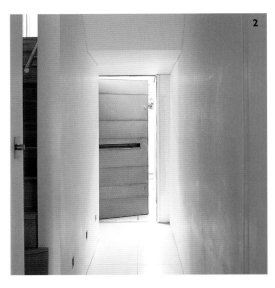

1 The new stair tower is sided with copper sheet. This area was once the side alley to the existing house. **2** The new stairwell has been relocated to the original position of the side alley of the old house, using this otherwise useless space and allowing for complete open levels within the new floor plan. The pivoting copper covered door opens onto a paved outdoor area.

Besides creating a modern home, this project—by architectural firm Curtis Wood—has entailed renovating a historic façade, reinstating missing period details, constructing a new building, and extending the original floor plan.

Located on the riverbank, this historic house had been lived in by the previous owners for many years. It was once a warren of small rooms with leaking roofs and even boasted its own elevator shaft. The idea to tear down the old house and rebuild behind the original façade was the best option; not only was it in bad condition, but working this way would save extra expense due to the fact that no tax would be payable on such a refurbishment.

The idea to make each floor as open as possible came together when the decision was made to build a stair tower away from the main part of the house. This tower would be positioned in the original alleyway at the side of the building. The copper siding on the tower would identify it as a new addition to the existing house, and would be the embelishment to the otherwise unadorned home, except for the features on the turn-of-the-century façade.

Inspired by Pawson's work, the owners of this house did not want the interior to be too clinical. The materials chosen—limestone for the ground floor, oak for the other floors and staircase, and walnut for all the built-in furniture—are complemented with the owners' interest in modern furniture. In the main living area situated on the first floor, there are large leather sofas from the 1970s, a pair of 1950s Italian armchairs upholstered in velvet, and the occasional freestanding floor light; while much of the lighting is concealed, the architects realised the positive effect of freestanding lighting in these areas.

The new house behind the façade consists of three floors of accommodation with a large roof terrace, accessed by a continuation of the main staircase within the copper clad tower. The first floor consists of the main living area, the kitchen

3 In the kitchen John Pawson's idea of keeping everything out of sight is adhered to. This wall of storage houses all the kitchen appliances, utensils, and dishes. The ground floor, now rebuilt, spans completely from the front through to the rear of the house. A new structural system had to be installed to allow for such open areas. The open-plan sitting area leads into the kitchen space, and beyond to the dining zone. An open fireplace runs along the left-hand side of the sitting area. The stairwell is just visible. The furniture is vintage mid-century.

4 From the galley-type kitchen, the open plan space leads to the dining area. The glass wall can be completely opened, enlarging this area to the boundary wall of the backyard. **5** The accordion doors run along a concealed track and fold against the interior wall. There is one slender support at the corner that supports above, enabling all the glass to be opened. The concealed curved track is for the drapery.

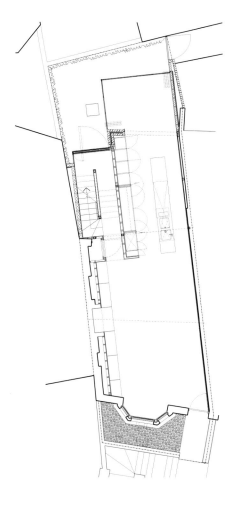

FLOOR PLAN ABOVE
The new open-plan floor plan shows the new location of the staircase. The kitchen wall storage divides the space, while leading your eye through to the dining area beyond.

space, and a dining extension that completely opens up to the outside area. A limestone floor carries on throughout the entire ground floor. On the second floor is a more formal living room with a full-length fireplace, and a walnut-faced built-in area that conceals the entertainment equipment, a bathroom, and a small bedroom.

The top floor contains the main bedroom with an entire wall of walnut storage, a shower room with glass roof *à la* Pawson, and another bedroom. Throughout the house,

ideas and principles from Pawson's work can be noted—"the idea to hide all," the concealed lighting, the slab-like kitchen island, and low-level bench along the fireplace wall.

This is an extreme way to reorganize space within your home. It shows the importance of retaining the original façade within a row of houses. It works on the principle of relocating the staircase to open up the floor plan. With its mix of hard and soft materials, a timeless modern home has been created that will stand the test of time.

TWO INTO ONE

What were originally two small apartments have been united to produce an attractive and practical duplex. The two fifty-square-meter spaces have been combined to create a modern and compact home.

This approach to creating the ideal home entails stripping bare the original spaces and installing a completely new layout with what you want where you want it. As this space was originally two units, it had two kitchens and two bathrooms. Built around 1900, the building was one of the first residential developments of the extension of Amsterdam's inner city. Now it is considered as being in the very heart of the city.

The two spaces are pulled together by the installation of the new steel staircase that is placed in the center of the units and breaks through from one floor to the next. Originally this area was the sleeping zone of the lower space. A large three-square-meter void has been left in this position, and creates a massive volume of space.

This now large one-bedroom duplex has its main relaxing space on the upper level opposite from the bedroom area. Back on the lower level, as you enter you come across a wall of translucent storage that in turn, opens into a large bathroom that is conveniently positioned in the footprint of the previous one.

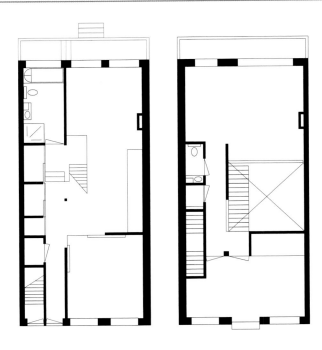

GROUND FLOOR ABOVE LEFT
What were two apartments are now one duplex. The ground floor was originally a long railroad type space divided by a glazed-in bedroom area. The second-floor apartment was partitioned into two spaces.

SECOND FLOOR ABOVE RIGHT
The installation of the central staircase unites the two levels; the kitchen, dining, and work areas are on the first floor while the bedroom and main living area is on the second.

1 From the entrance area the staircase screens off the kitchen area, and draws your eye to the dining space with its full-height French doors leading to the terrace. **2** The simple white kitchen runs the length of the central area of the duplex. Using this stairway area also as a kitchen makes the most of all the space in this home. **3** The steel-framed wood-tread staircase zigzags up to the bedroom and main living area. Behind the staircase on the first floor lies the study with its full-height sliding door. The double-height partition wall is faced with figured veneer plywood. **4** The partition wall rises from below, and screens off the bedroom area across the void of the staircase and the kitchen below. The classic pieces of furniture include a tubular frame table by Marcel Breuer and a Barcelona chair by Mies van der Rohe.

The kitchen is located in the center of the overall space opposite the entrance area, and leads into the dining area, which includes a traditional approach to the fireplace. From the dining area, French doors lead onto a small terrace. These doors and windows add proportion and order to the house's original framework. An open-tread staircase keeps the space free flowing, and appears to hover within the void. With its simple zigzag steel profile structure, it rises from the elevated platform to the living and sleeping areas.

One of the few walls left standing carries on up from the lower to the upper levels. It is plywood faced and acts as the main division of the space. A study and bedroom are placed behind it. Both rooms also have full-height sliding doors to close off these areas when necessary. The living/relaxing zone on the second floor takes you away from the kitchen and dining area below, and utilizes the space available to its best advantage. Today, more social areas are taking preference over large bedrooms and bathrooms. The open-plan approach of this space means that the living areas are both accessible and highly usable. With ample storage in the entrance foyer, the entire space is clutter-free and streamlined.

The materials used—oak floors, steel staircase, and white painted walls, polycarbonate sheeting for the storage area, and the figured plywood for the partition wall, create a simple overall united space.

STREAMLINE MODERN

This 19th century depot, formerly storage for horse and carriage and trams, was converted into a school in the 1920s, and, after being initially developed into private apartments in the mid 1990s, has more recently been transformed by architect Marc Prosman.

Original features from its previous life are evident on the first floor of this building. The massive wooden beams date from the initial depot's construction. These huge supports give scale to the entire space, and show how one can successfully balance the old with the new, when developing an older building; it is important to create visual harmony. Using original structures as dividing elements of a space both highlights and incorporates them into the new scheme, while avoiding restricting the overall interior, and without allowing them to dominate the interior.

The previous layout consisted of a mishmash of unusual shaped rooms with diagonal walls and narrow openings.

2

1 From the central lobby, the screen-printed double swing doors open onto the lower level of the living area. The travertine floor carries on through. The high-level windows above the kitchen area give ample light, without the worry of being overlooked.
2 Looking from the raised sitting area, the study area is visible. The full-height veneered partition is the identical thickness to the main wooden structure beam. The kitchen elements, designed by John Pawson, appear to be movable within the space. The window at the end overlooks the apartment's garden.

Upon entering the apartment, one comes to the bedroom and bathroom areas. All these spaces have been reorganized into streamlined rooms, each kitted out with ample storage and bathroom areas. Keeping all the main services in their original places speeds up the renovation process, as you do not need to lay new pipework and plumbing facilities.

From the inner foyer, double silk screened glass doors lead into the main living space.

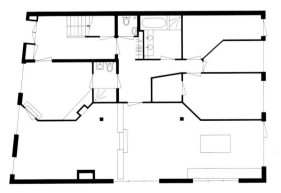

BEFORE FAR LEFT
A warren of strangely angled rooms with what seem to be rooms within rooms. Original features such as columns are well exposed.

AFTER LEFT
With most internal walls either moved or removed, the new plan is streamlined. The kitchen and bathrooms remain where they were, enabling an easier and quicker renovation. With the use of sliding partitions, every area is usable, but can be enclosed when necessary.

3 The space in this apartment has been juggled around to make a home that is both flexible and precise and that is a timeless in modern design. Upon entering the living space from the interior hallway, the openness of the apartment is apparent. Almost every element appears to be moveable. The desk chair on wheels is by Arne Jacobsen. **4** The track of the pocket door is hung from the original structural beam. This door slides into the dividing partition between the kitchen and the study areas. **5** In the study, a custom-made desk makes the ideal work station. There is a pocket door within the full-height partition. The massive beam runs right through the apartment.

5

Utilizing the existing raised level in the main part of the apartment adds another dimension to the relaxing zone. With the cooking, eating, and relaxing areas comprising an open-plan layout, the entire space is treated as a whole. The kitchen has been designed by John Pawson for this apartment, and carries on the simple solid design elements of the interior. With white marble in the bathrooms, wood for the dividing screens and walls, and travertine for the floors, these elaborate details bring a fresh dimension to the otherwise traditional materials. Across from the kitchen area is the study area that leads to one of the bedrooms. The dividing wall to the study area is a wood-paneled structure that is the same thickness as the massive structural beams. It also conceals a pocket door so this area can be closed off when necessary. A partition divides the study from the bedroom beyond. This storage wall houses all work necessities on the study side, while on the bedroom side, it holds a plasma screen.

Almost every wall of this apartment has a dual purpose, providing either a screen from another area, a wall of storage or cabinets, or full-height sliding or pivoting doors. This enables the apartment to be opened up, and for one space to lead seamlessly to another, while also creating intimate zones for a modern home.

This end of the space has two Vladimir Kagan-designed daybeds, a grouping of Noguchi lamps, and two Brancusi-inspired ceramic tables. Nothing is built in, so all the items are movable to allow the space to be used for other functions, for example, a sleepover. With the screens entirely closed, this area becomes a translucent box within the space.

The screens slide and fold, and when not in use, park against the wall in an alcove. The system glides smoothly and makes for easy and quick partitioning. They can be left in various configurations to suit the occasion.

This enclosed space has windows onto the front paved yard. The original shutters and ironwork recall the age and style of the townhouse. The basement still has access to the street; even better for guests, who can come and go without disturbing the main household.

Reorganizing the space you have in your home allows you to figure out what you really need, what you want where, and means you can open up unused spaces to provide access to every square foot of your home. It also lets you reassess your total living space, and enables you to turn your existing home into a modern living experience for you and your family.

In the basement of this house, extended and reorganized by the firm Ogawa Depardon, the owners wanted to create a family room for the children to play in and in which the family could relax. It is also the main eating area and offers enough space at the front of the house to accommodate the occasional guest.

Once a rental unit with a network of rooms, the entire space has been opened up. The central part has a sinuous deSede sectional sofa, a display area for favorite artworks, and oak-paneled built-in storage for the plasma screen T, and all its accessories.

The flexibility of this floorplan is what makes it work so well. There is plenty of room for the entire family to be able to relax, work, or play, all in the same area. The fact that the floor also opens out onto the backyard only adds to its success. The ingenious installation of the folding screens at the front of the house gives a further dimension to this basement.

1 With the screens closed, a great white light floods the room. The curving deSede sofa takes away the railroad effect the room could have had. On the right is the storage for the entertainment equipment. **2** A structural support acts as the end stop for the sliding screens. With the tracks recessed when the screens are open there is no visual break in the overall ceiling plane. The Noguchi lamps complement the ceramic table. **3** With the screens half open, this area looks inviting yet enclosed. It could be a quiet area to study, read, or play the guitar. The Plexiglas legs of the daybeds give the impression of floating within the space.

SPACIOUS FAMILY LOFT

FLOORPLAN RIGHT
The plan showing the reorganization of the loft space appears to be a free-flowing circular solution with a central hub of the kitchen and bathroom around the existing building's staircase. The children's rooms are allocated to the angled rooms, away from the parents' zone. An elevator opens into the apartment's lobby area.

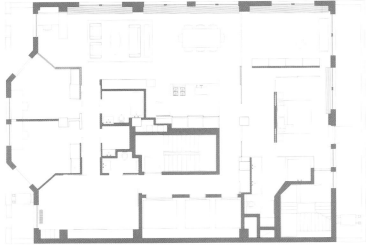

An urban loft has been reorganized for the use of a conventional family set up: mom, dad, and their two children.

After moving from a smaller space in a similar area, the owners Jo Shane and John Cooper brought with them a collection of family hand-me-downs, consisting of important design pieces—ranging from Eero Saarinen's womb chair, to George Nelson's coconut chairs. These pieces needed their own space, and prompted the overall direction of the interior.

Taking the 2,500 square foot unit, the firm 1100 Architect have created a unique modern space; while including the industrial elements such as the beams, the sprinkler systems, and the original concrete floors, they have added slick new features such as the wide walnut floorboards, the sliding translucent doors and partitions, and the use of wood-effect laminates in the utility areas, including the kitchen.

The whole area is split up into various elements. The central core includes the open-plan kitchen, which is flanked with storage behind opaque glass screens (that can be illuminated at night), and pull-out units in the faux walnut wall that back onto the children's wing.

On the west side of the apartment, a complete wall of windows houses the den area—raised slightly off the original concrete slab—comprising a dining area, a living area, and a small home office. The different level and flooring materials make this room feel more enclosed, and invitingly intimate.

1 With the screens to the den closed, the dining area feels more enclosed and the den becomes a separate unit within the entire space. The George Nelson crossed-leg dining table and the Harry Bertoia dining chairs give a solid design statement to the overall space. The Nelson-inspired bench creates a sense of proportion. 2 With the open screens offering access to the den, the area becomes welcoming and useful. The slightly raised floor level keeps it within its boundaries, yet makes for an inviting space. The wall of storage houses the book collection. The Saarinen womb chair and rare ottoman are both family heirlooms.

3 With its concrete countertop, the kitchen island acts as the hub of the apartment. The Bertoia barstools are solid yet see-through. The translucent screens on the back wall slide away to expose ample shelving for the useful and precious collection of pots and pans. The faux walnut wall includes pull-out storage and flip-down work areas. The hanging lights are by Biproduct.

4 The master bedroom faces north. The slatted storage bench runs along this room, as in the living areas. The sliding storage area offers yet more storage, this time for clothes. At a high level, the sprinkler system is well exposed. **5** The dressing room backs onto the master bedroom. The hanging clothes create an ever-changing color backdrop in this area. The fact that this area is on a raised floor gives it feelings of suspension and movability.

5

4

Along the entire length of this open living, dining, and working space runs a window bench that also works as a necessary storage unit.

The master bedroom is placed behind the dressing room that, in addition to being the walk-in closet, brings light to the walkway, from the main entrance of the apartment to the living areas. This bedroom, on the north side of the building, includes a connecting shower room and has access to a small terrace. The space here has been planned so that the children have their own spaces which are detached from those of the adults who also have their own bathroom.

This loft apartment also includes an art studio with direct access to the elevator and to the living accommodation. This working zone can be included in the whole space seamlessly since it does not interfere with the overall communal living areas.

The reorganization of space in this house shows, with a balance of the original elements—the introduction of new finishes, the undertaking of the overall demands of the owners, and the use of existing furniture and furnishing— how easy, flowing and comfortable a space like this can be.

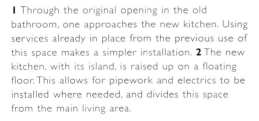

1 Through the original opening in the old bathroom, one approaches the new kitchen. Using services already in place from the previous use of this space makes a simpler installation. **2** The new kitchen, with its island, is raised up on a floating floor. This allows for pipework and electrics to be installed where needed, and divides this space from the main living area.

In this ex-industrial space the entrance level has been reorganized to make the place more usable. Originally, a small lobby opened onto an internal hall with more doorways, one to an enclosed kitchen, and the other to a bathroom. The bathroom has been relocated to the bedroom floor, while the kitchen has been opened up to the main living area, and is in the footprint of the old bathroom. The idea of loft living is to take advantage of an open-plan layout of spaces as opposed to the conventional configuration of individual self-contained rooms. Previously, this property was a live/work unit, and that is why the rooms were more enclosed. Now completely residential, the entire space needed to be opened up and usable for day-to-day activities, and to create a more sociable and entertaining environment.

3 The dining area is in the place of the old kitchen. With its double-height volume, it is flooded with light from the skylight above. A wall of storage adds visual impact. **4** A useful downstairs bathroom has been placed in the original entrance to the old kitchen. This area combines storage and display, sometimes forgotten about in a practical room of this kind.

To allow heating ductwork to get to opposite sides of the entire space, and to get pipework to the central kitchen island and the downstairs powder room, the floor has been raised in the new kitchen, dining area, and hall. This actually adds to the entire layout of the loft as it creates a slightly lower area for the main reception and den areas. It also allows for different floor surfaces, which in turn provide visually effective divisions of the space.

In the original entrance to the old kitchen, a small powder room has been installed. Using new technology for waste disposal, small diameter pipework could be used for this area. Within the old kitchen space, the new dining area is surrounded with floor-to-ceiling storage. Using chain-store units, the doors are simply painted board, cut to size as required, creating a visual and useful area. The central kitchen island is made from basic units topped with

3

a white Corian work surface with an inset sink, and a dishwasher and a washing machine housed underneath.

The all-white decoration helps blend the old brickwork with the new plasterwork. A massive structural brick wall running the length of the space divides the area, keeping the location of certain elements obvious, such as the kitchen and bathroom, due to the positions of openings and services.

In this kind of space, visual elements are important and useful. Due to the almost 50-foot vista, the grid of storage at one end relates well to the grid of the window at the other end, bringing the two ends together.

EXTENDING YOUR SPACE

Searching for that perfect modern space can be a lengthy and often unrewarding undertaking. If you need more space, or if you cannot find the right sized accommodation in the area in which you want to live, instead of going through the trauma of moving, extending your space could prove the best alternative. Making your space bigger is becoming increasingly the way forward in adding a modern element to your existing or new space.

If you like where you live, but are short of space, look at what you have and revalue all your options. You can go upward, downward or outward. You can perhaps build on the roof, utilize the attic space, make use of the basement, or add on to the rear or side elevation.

Of course, all these options need careful consideration and lengthy conversations with your local building office. It is also a good idea to search for specialized firms and consult them about what you want before you get more involved, thereby helping you to calculate whether your ideas will actually come within your budget. Since these projects are usually mainly structural, the appropriate people need to be involved from the outset, to check that your proposals are feasible. Such developments do not usually need planning permission but most do need code permits and control.

First consider the potential of your roof space; with a typical pitched roof you can use the space inside the roof or extend out, or in some cases, raise the roof line. In a row of houses, if it is evident that some of the properties have roof

extensions, it is usually a telltale sign that you can do the same. In historic areas, any extension will usually have to be in a similar style and built using similar materials to those of surrounding properties of a similar period. However, in certain areas, some authorities allow totally modern additions. In an area of local development, you can sometimes be more adventurous with your scheme, using different materials and can even include balconies and open rooftop spaces. Sometimes developing within the limits of an existing roof space can be limiting. An alternative option is to use the roof space to simply increase the volume of rooms beneath.

If your house has a flat roof or if you live in the penthouse level of an apartment building, you may be able to build directly on top of your existing space. This can prove a lengthy and expensive project, but it can double your apartment's size.

Your basement could also be ripe for development, particularly if the ceiling height is already adequate. Digging deeper is possible, and this option will give you ample ceiling height; however, the whole space would need to be tanked out to stop the possibility of the water level encroaching the space. This basement space is usually ideal for accommodating a den or entertainment room, or for creating a kitchen/dining area, or even a workshop area, as it is away from the main part of the house and could be treated in a completely different decorative style to the main part of the house. If your house has a raised entrance, the basement may have access to the backyard, therefore giving important access between the indoor and outdoor space.

An addition is perhaps the easiest and least disruptive way of extending your space. If your yard is big enough, you can extend quite far. With so many products available now for doors and windows, complete glass spaces can be built, but privacy is a factor that should not be overlooked when planning a glassed-in addition.

Of course, the original spaces that are the openings to any new addition need to blend in with the new. There needs to be a natural progression to the new space from the old.

1 The extension of glass panel construction backs onto a 1930s Modernist house. The curves complement the bay window on the front elevation, and the curved utility area, now the kitchen, at the rear. 2 On the roof terrace of a river-side house, the staircase opens up from the copper-covered box. The outdoor kitchen on the terrace saves the residents from running down three flights of stairs when using the area.

SIGNIFICANT IMPROVEMENTS

In this house built by Don Chapell,

the space has been enlarged more than once. Initially, this quite small house of 2800 square feet was built and lived in by the architect during his bachelor days. Once the present owners, Lois and Les Fishman, moved in, they realized the kitchen area was not going to be big enough. So they called in the cabinetmaker, Dale Rieke, to expand the kitchen and eating areas. They pushed out the front elevation wall to a slight curve, and raised the roof level to give greater volume to the dining area. When you are in the new kitchen, you are not aware of the curve of the new elevation, but it is

1 The kitchen has been extended visibly by the slightly curving elevation above the parking spaces. Behind is the raised roof line of the dining area. **2** Typical of Chapell's work, the kitchen color scheme relates to the landscape seen through the windows.

3 The new dining area has a raised roof that allows light to flood in through through its vertical and horizontal arrangement of windows. The painting over the dining area inspired the color scheme in this space.

clearly visible from the exterior. The irregularity of the cabinets is visually exciting.and the subtle color palette of units relates to the tones and hues found in the surrounding landscape. The central island gives ample work surface for food preparation and entertaining.

The owners convinced Chapell to sell them the adjacent lot, to make sure there was no possibility of another house being built.

4 The corner glass doors that lead onto the terrace slide into pockets when open, allowing for a complete inside-outside lifestyle. The purple womb chair brings in the landscape beyond. 5 The existing living room has an opening to allow views of the new room beyond. This divides the areas, but also entices you to experience the new space.

The Fishmans had planned for Chapell to extend the house and add a wing to it. But after his untimely death, they called in Guy Peterson. A follower of work by Paul Rudolph, as was Chapell, Peterson seemed the likely candidate.

This new wing was to almost double the volume of the existing house. Peterson suggested removing the side wall of the house, not to give access, but to open up the visual aspect of the new living room beyond. The Fishmans were not sure about this, but they soon realized it was one of the most significant improvements to their living space.

The existing living space, with built-in furniture by Lois's son Bill Mostow, also includes 20th century classics. In the new living space, more classics tie this area to the existing interior style. Wilson Stiles,

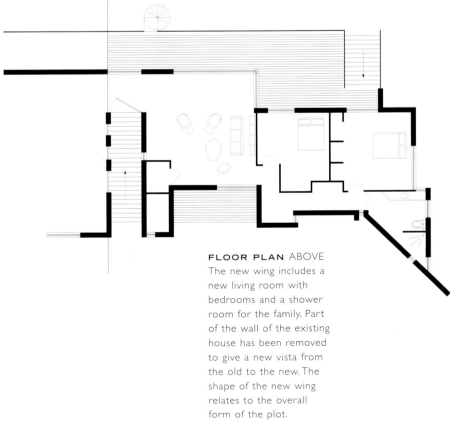

FLOOR PLAN ABOVE
The new wing includes a new living room with bedrooms and a shower room for the family. Part of the wall of the existing house has been removed to give a new vista from the old to the new. The shape of the new wing relates to the overall form of the plot.

responsible for the interior design, has used color from the landscape to successfully bring the outside in, as in Chapell's kitchen design. The Saarinen womb chair matches the bougainvilea nearby.

At the far end of the house, a projecting triangular wing outlines the position of the new shower room. This angled space relates to the lot's shape, and brings the house to a halt. Light pouring in from the vertical floor-to-ceiling slit window gives a feeling of infinity, and links the colored mosaic tiled walls to the natural seascape outside. As Les Fishman says, "We have bright vistas from room to room, and every room has a view."

6 The triangular projection at the end of the house relates to the shape of the lot following the coastal line. Within the tapered walls, a large shower room is housed. **7** Built-in cabinetry is angled to fit the irregular space of this area. **8** The tiles used in the shower area relate to the landscape outside.

This former warehouse

building has been carefully converted into a dynamic and spacious residential loft.

Having lived in the Tribeca neigborhood of New York, for many years, the owners of this space joined forces with their neighbors to try to find a suitable building to convert for themselves. They found a seven-story building that was built in the early 1900s. It still retained many of the typical loft features, such as the bare brick walls and huge wooden rafters. Between them, they calculated they would share the top two floors.

Working with architect firm The Downtown Group, they called in Mark Winkelman to calculate the division of space. After various considerations, they decided to divide the two floors with a diagonal wall, creating a wedge-shaped unit that is about 6 feet at one end and 50 feet at the other. Working on the plans was a bit of a problem. The wedge-shaped plan creates a long widening entry to the space. To get the best south-facing views, the architects came up with the design for a new floor on top of the existing building. The main living room would be in the new rooftop extension, which would overlook the entrance hall. This meant the kitchen, dining area, study, master bedroom, and bathroom would be on the seventh floor, the entry floor. The children would

1 From the roof terrace, the roof line is more apparent. The wall on the left is the dividing line between the two lofts. To the right is the cutout for the courtyard below. 2 Above the staircase, the roof lights bring light right down to the children's level on the lower floor. These thin metal beams appear to be keeping the roof under control. 3 From the new rooftop floor, the cedar-paneled hyper-parabolic roof sweeps across the living area. The wall of glass doors opens onto a large roof terrace.

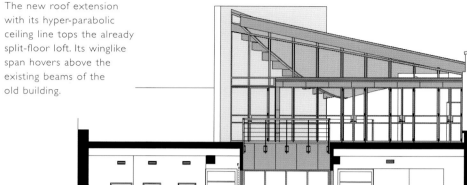

AFTER RIGHT
The new roof extension with its hyper-parabolic ceiling line tops the already split-floor loft. Its winglike span hovers above the existing beams of the old building.

have their own space down on the sixth floor. There would also be an open courtyard on the main floor, with three sides opening.

Using a computer model, Winkelman and Valledor, the project designers, were able to work out the precise angles for the new roof line. A cedar-paneled hyper-parabolic ceiling and roof was designed. This roof line would sweep up like a wing above the existing heavy beams of the old building.

The steel staircase lifts up from the entrance area, under the wooden supports of the seventh floor, and is naturally lit from above by the glazed panels that run along the edge of the roof at the dividing diagonal wall.

Throughout the entire loft, well-designed cherrywood built-in storage and kitchen cabinets have been installed. With the study leading off the kitchen and into the courtyard, this outdoor space can also become an informal breakfast area whenever it is desired. At the roof level, an entire wall of metal-framed doors and windows can be opened for access to the huge roof terrace. With a cutout for the courtyard below, this outdoor space works well for entertaining.

This combination of split levels and exposure to ample daylight has given the converted loft space much more capacity and a dramatic feel. With so many flat roofs around, you can imagine this could be the best way to increase the size of your house without moving; but remember, when you are working on a project of this scale, architects and engineers have to be consulted.

Space that is already part of a house, but not included in the living areas, is often overlooked. These days, having a garage is a luxury, but with many older properties, a standard garage is just not big enough for today's average-size car or 4 x 4 vehicle. Whether it is part of the main structure or an original addition to the side, these integral spaces are ideal extensions to the existing living space.

In this house, built in 1935, the garage is just under eight feet wide, and with the doors open, the opening narrows to just over seven feet. Garages today seem to be more for storage: the more space you have, the more you need. Luckily, with this house, the decision to develop the garage space was made while the whole house was being worked on.

Originally, this home had no access to the large backyard apart from a back door from the kitchen. Even the rooms upstairs did not have access to the flat roof at the rear. Strangely, there are two balconies with doors opening onto them at the front of the house, overlooking the road. However, with well-positioned trees and more planting, these are now more private.

While bricked-up openings were cut out of the end wall of the garage, the back wall of the room above was removed to accommodate French doors, which let in great views of

1 Looking out from the new summer room through the sliding doors, one can enjoy a great view of the swimming pool. This room, for relaxing and summer eating, looks like it has always been part of the house, even though it was originally the garage and then converted to its current use. Even in the winter with the doors closed, it is still a great space in which to relax. The dropped ceiling has been painted to complement the pool and exaggerate the blue skies of summer. 2 The back elevation of the house shows how the balcony and large space beyond the old garage gives space for lounging and access to the pool. The door and railings above appear to have always been there.

BEFORE ABOVE LEFT
The cramped entrance hall and group of small rooms around the kitchen area make the ground floor very traditional compared to the very modern exterior. The original garage space takes up a great deal of the floorplan, ideal to convert.

AFTER ABOVE RIGHT
The garage doors have been replaced with a solid wall. The partition walls have been removed, giving an open plan approach to the kitchen and entrance hall. Double doors open onto the back garden.

3 The two porthole windows were inserted into the wall that was originally the entrance to the garage. These were salvaged from other areas within the house. **4** From the original living room you can see into the new summer room and out of the French doors where the kitchen was originally located. The central staircase has been stripped of paint to carry the wood flooring up to the next level. **5** The downstairs powder room was made slightly smaller to allow for an opening to the new room. The circular window theme carries through.

the swimming pool and sea beyond. An opening from the hall was made in the old garage and to accomplish this, the large powder room under the staircase had to be shortened. One can now look from the existing living room into the new summer room.

Large sliding doors, similar to those used in automobile showrooms, were installed on the ground level at the end of the garage, so that almost all of it could be open during the summer months. In the room above, Crittall metal-framed French doors were installed, and a tubular nautical railing was installed around the balcony area.

The old garage had two porthole windows on the side that now add to the overall nautical theme. Inside the new room, reclaimed woodblock flooring was sourced and laid with a staggered joint to match the kitchen area. Two more porthole windows were installed in the wall that replaced the double doors. The theme from other parts of the house is carried on throughout, primarily using furniture by Alvar Aalto.

Elsewhere in the house the ground floor was opened up. There was originally a warren of rooms in the central core of the house. A large walk-in coat closet, the kitchen, a bathroom, and a maid's room were all removed to make an open-plan space to house the new kitchen. French doors, the same as those in the room above the summer room, were installed into the same position as the original kitchen window, opening up this area to the backyard as well.

In all the ground floor rooms with openings, the inside and outside flooring materials have been laid as near to the same levels as possible to create an uninterrupted plane, so that one area flows into the next.

This extension of space shows how important it is to retain features and specifications from the existing house and structure. To use new versions of the windows, but keeping to the original proportions, is one of the most important elements to consider when working on an older building. Carrying on the same or similar flooring materials unites the old with the new.

This apartment has been extended by laterally taking over the space in the adjoining building. The original studio apartment on the Upper West Side of Manhattan, with studio and kitchen on one floor and a bedroom and bathroom on the upper floor, was developed two years previously. New materials, both natural and synthetic, were used. Refurbished vintage steel cabinets have been used in the kitchen, alongside the natural olive wood used for the new cabinets and shelving. Concrete has been used for the countertops with blackened steel for the fireplace.

The chance to extend laterally is unusual. In this case, as the two townhouses are in fact part of the same coop ownership, they were seen as one property in the eyes of the building department. It helped a great deal that the floor levels were the same.

In the adjacent townhouse, there is a living room, bathroom, bedroom, and a small office area. In this area, more synthetic materials have been used. In the bathroom, there is a resin and pebble slab floor, and fiberglass sliding panels open up and close off certain areas.

The apartment now has two bathrooms. The original one on the top floor is more traditional in approach with its mosaic tile and stone floor, while the new bathroom is more up-to-date. This space is treated as an open shower room with a slot in the floor for drainage.

The total decoration of the apartment, combining vintage and oriental objects and furnishings, has a calming and timeless ambience. The use of color, texture, and materials has created an eclectic mix of style and culture. From the Bertoia breakfast stools to the trolley car used as a table, each individual item works well with the next. Both apartment renovations are by Ogawa Depardon and are united by a roof terrace, accessed from either unit via custom-made pivoting doors.

In this kind of extension, taking over the space next to you is an ambitious way to gain more accommodation. It is perhaps an extravagant way of extending your home space, but in reality it is somewhat similar to converting your basement or attic. Initially, you will end up with two of this and two of that, however, it will increase your living space to at least double its size without your having to move.

FLOOR PLAN RIGHT
What was once a duplex on one side and a railroad apartment on the other have been combined, creating a double drawing room, study area and more private living room at the front of both buildings. Outside, the roof terrace becomes an outdoor room.

1 This is the kitchen in the original studio apartment with bathroom and bedroom upstairs. Refurbished steel cabinets have been used in this zone. The stained oak staircase wraps around the kitchen and continues the dark floor upstairs. **2** From the second-phase renovation you can see the original studio. This space in the adjoining townhouse has a bleached oak floor while the original apartment has a black ink-stained finish.

GLAZED LEVELS UNITED

Upon entering this five-story Victorian

townhouse in Brooklyn, New York, one has no idea what is found at the back of the house. The front elevation retains all of its original features—doors, windows, and ironwork—but once the front door is opened, it is clear that there is too much light filtering through the engraved glass doors for this to be an entirely original house. What would have been a dark passage to the original kitchen is now a floodlit entrance to a double-height atrium, housing the kitchen on the second floor and the dining area below. The New York-based firm Ogawa Depardon has added one of its modern signature pieces to an otherwise aged, traditional home. This two-story addition has united the former maze of tiny rooms on the lower floor with the space on the entrance level.

Through what must have been the French doors out of the main reception room onto a terrace is now the new kitchen. With a typical Ogawa Depardon ceiling-hung unit dividing the space, the kitchen is open to the room below and contains views of the garden beyond. The new staircase flows naturally down to the new entertaining floor that houses the dining room, and offers one large den space that can be divided by sliding the glass partitions. Leaving these partition doors open facilitates the use of the entire lower level, making it a focal area for socializing.

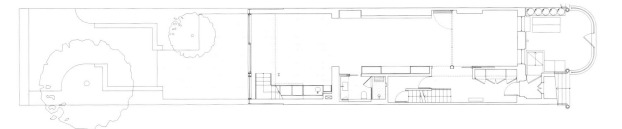

BEFORE LEFT
The original layout of the lower level was a series of very dark rooms. This was originally a self-contained unit that housed its own kitchen and bathroom.

AFTER BELOW LEFT
Now the entire lower level has been opened up to house a shower room and wet area, and these now replace the previous areas. The removal of the rear and dividing walls has managed to visually create more space.

2

I The double-height extension on the rear of this Victorian house replaces a smaller lower-level box room and opens up the living areas of the house. The sliding doors bring the outside in and, while modern in construction, still retain the three-window grid of the original house elevation. **2** An aesthetically-pleasing steel frame creates the grid for the new windows. The thinness of the edge of the steel beams gives a lightness to the overall mass. **3** The dining area at the lower level opens onto the garden. The stairway from the kitchen above sweeps down into the space. The interior, created by Michael Formica, brings color into an area that could easily be dull and dark.

BEFORE ABOVE
The rear of the house was originally a mishmash of windows and doors with a small rear addition and an outside staircase. There was already a terrace leading outside from the second-story rooms.

AFTER BELOW
The new addition echoes the three window grid of the floors above, and dramatically increases the amount of light that floods the lower floors.

3

4 In the kitchen, the bank of cabinets houses the dishwasher and the sink. The view here, over the lower level and into the yard, could not be any better. From this level, the stairs lead directly to the dining area below, forming a natural flow from the one floor to another. **5** From what must have once been the original opening at the back of the house, the new kitchen space comes into view. The ceiling-hung storage unit divides the space and balances the grid of the windows beyond. The matching floor from the old to the new unites the space—an important element often overlooked when extending your space.

BEFORE ABOVE RIGHT
The lower level before was basically two rooms with a kitchen and bathroom, and represented a separate unit from the main house.

AFTER BELOW RIGHT
Now with the introduction of a staircase from the new kitchen, the lower level is now incorporated into the entire space of the house.

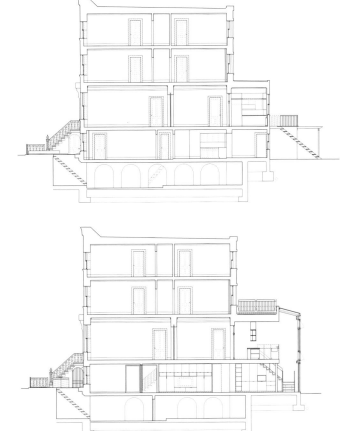

In some way, when doing an addition that houses the kitchen and dining area, one is tempted to put the kitchen down in the lowest part of the house. By placing this kitchen on the entrance level, it is more usable and accessible from the floors above. It also means that the entire lower floor can be given over to a general family room, thus keeping the main living room off the entrance hall as a quiet and calm space while retaining the original layout of the property.

From the backyard, the new elevation follows the original three-window grid of the rear of the house. It has a very simple cross-frame skeleton that carries the glazed units, both fixed and opening. The mix of steel and wood frames mellows the whole appearance and gives the feeling that it has been there a while, and which naturally works with the surroundings.

Overall, the addition to the house has been created for the family to use, so it is flexible and opens up otherwise useless spaces that one often finds in older homes like this.

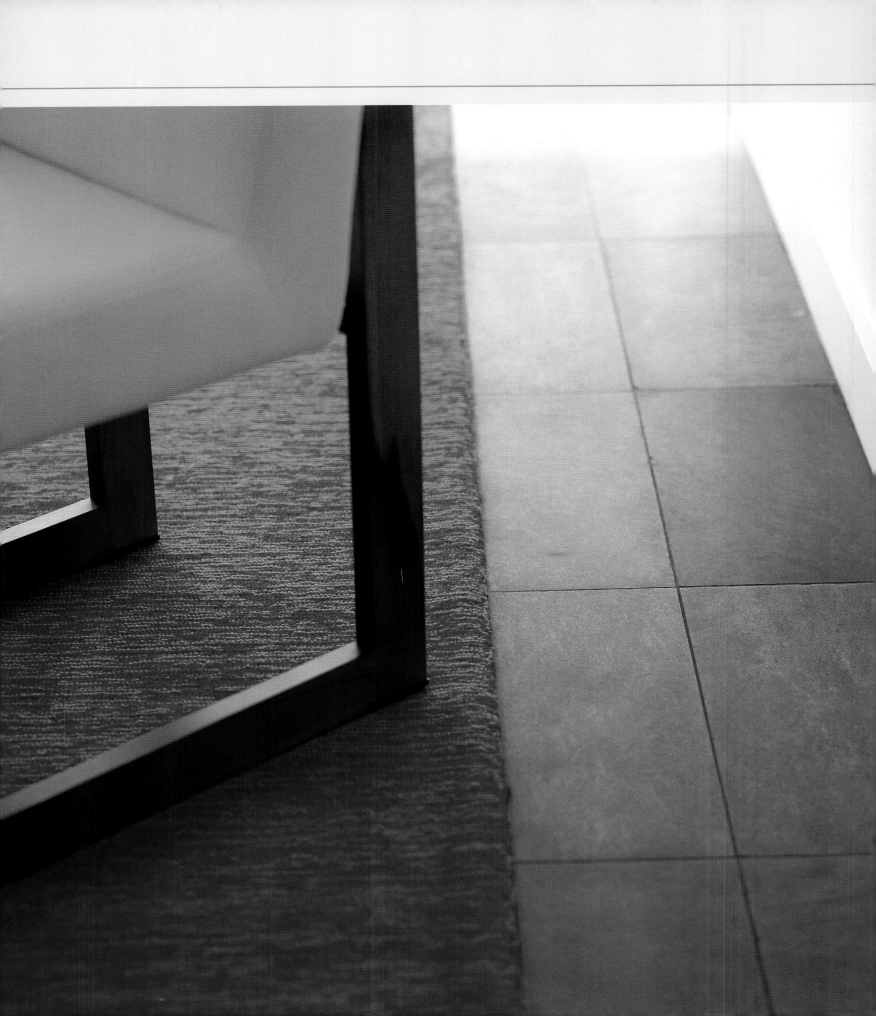

THE HOME DISSECTED

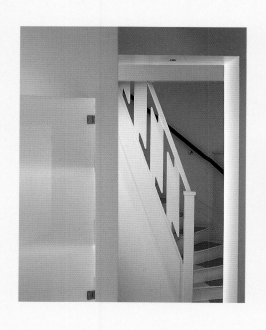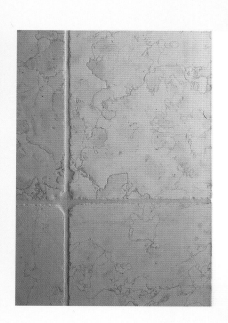

Page 92 Contrasting floor surfaces give a distinctive division to an open-plan space, at the same time adding textural contrast and clean visual lines. **1** Highlighting the original structural elements of this Amsterdam house gives uniformity and proportion to a modern home. **2** Again in Amsterdam, this wall of storage is divided up into smaller units marked by flat-fronted doors in various sizes. It is visually exciting while still being useful. The chair is by Arne Jacobsen.

Whether you have just bought a new home, apartment, or industrial unit, or decided that now is the time to update your existing space, where to start and what to do are the two main questions. This section breaks down the elements of the home and gives solutions to inspire you.

There are certain aspects that relate to all of the various degrees of modernization. But perhaps there are only a few elements that you want to update. When making your home more modern, you may decide just to work on the storage areas or just your floors. You do not have to do it all. By selecting what is for you, you can work out what you are aiming for and you can get some idea of what approach to take and what to do first.

Before you even consider each element, try to figure out what you want to achieve. Assess your personal requirements, those of your children if you have them, whether you work from home, and also how you will, of course, be entertaining your friends. Getting the work done in the correct order is one of the most important parts of the whole procedure.

If you are taking the decorative modernization route, you may be looking to install new flooring materials, wall coverings and paints, or new or updated furniture. With this approach, you will not be installing anything as major as new wood floors, but you could try some of the new easy-to-lay flooring panels. You could paint your floorboards, or simply lay new carpets. On the walls you can paint, paper, or apply some other finish appropriate to the situation.

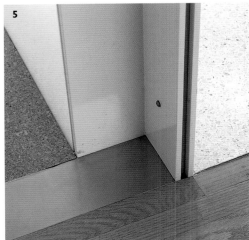

2

I would say, for the most impact, new furniture pieces for relaxing, storage, and display will change the home drastically. You can be more organized, show off your treasured items, and be able to put your feet up when all the work is done.

Updating more substantial features in the home will include laying more permanent new or reclaimed wood floors, ceramic-tile floors, or any other types of fixed flooring. You may want to change the doors to rooms, closets, and cupboards. In the kitchen you might update your existing cabinets with more modern and sleek faces. You may want to build in more unit furniture for display and storage, and you may make a new focal point in the main living area by updating the fireplace. You could replace the entire staircase—this is perhaps a

3 In the lobby of this Sarasota, Florida home a custom-built unit shows off a collection of European and Scandinavian ceramics. **4** These shelves designed by Alvar Aalto in the 1930s are still so modern. A combination of found objects and art are united in harmony. **5** A contrast of flooring materials here defines the spaces within the home, with terrazzo in the shower room, cork in the kitchen, and oak boards in the hall. The threshold strip is made of stainless steel.

6 Behind the zigzag staircase is a series of storage areas with doors of polycarbonate, giving a translucency to these areas. **7** A wall of walnut veneered doors runs the length of the house. Behind is ample storage for clothing.

little adventurous, but it will be a dramatic change. Of course lighting will play a part in updating your home. New hanging light fixtures or recessed spot lights can create great effects within the home without too much disruption.

If you are to reorganize your space, there are many more elements to consider. Maybe you will only be working on the main living and kitchen areas, but that could be enough. Think about uniting the space with completely new flooring throughout. In a bedroom and bathroom area, try using a duel-sided storage wall that would divide the two zones. Plan a lighting scheme that is flexible for many areas and uses—one that is dimmable and low energy, for example. Reorganizing your space gives you a chance to get your whole life sorted out, since you can include ample storage and that all-important work space.

If you are considering extending your home, every element will need to be considered. If you are going into the attic, for instance, you may need a new staircase. In other extended or added-on areas, you will need to unite the old with the new by using the same or similar flooring throughout. Other needs include lighting, and storage and display elements, and you may have a chance to include a new focal point to the extra space. If you are adding onto the back or side of the house, you will need to select windows and doors as well.

Whatever type of modernization you go for, lighting and furniture are the key visual elements. The lighting has to illuminate the room when necessary and also create atmosphere and ambience. The furniture has to be comfortable and do its job, but it also has to look good as it will dictate the feel of the space. Use either new design, vintage or reproduction classics or thrift-shop finds—it is all up to you—but they must work together to create the ideal modern home.

FLOORS

A good floor can make or

break a great interior. Whether you prefer floorboards—painted, sanded, or polished—hardwood strips, plywood panels, concrete, tile, or rubber, the choices are seemingly endless.

Working with what you already have is the best way to start. With good basic pine floorboards, you can sand and polish them, stain them, paint them, or you can use them as a solid surface for applying a multitude of other floor treatments. As long as it is well secured and pretty level, the wood floor can be a background material for many more decorative materials, especially other woods and various sheet floorings, such as birch plywood tongue-and-groove panels or more unusual finishes such as bamboo.

1 On the staircase and landing an oak plank floor shows the irregularity of the area. The walls appear to float as the flooring carries on beneath. **2** This plank floor runs throughout the apartment. Beyond these sliding panels is a Woodnotes rug that defines this bedroom. **3** This floor is a good example of using a semi-gloss sealer to preserve the wood's natural appearance. **4** This painted wood-strip floor shows how easy it is to update an otherwise tired and worn-out one. **5** Birch-faced plywood tongue-and-groove panels slot together to make an instant new floor. This flooring from Finland now comes prefinished, to make it even easier to install.

6 Using the same material for the treads of a staircase and for the landing areas maintains the flow. Here laminated walnut slabs are oiled for their protection.

7 A combination of the dark wood and the lighter limestone works in this totally new interior. The softness of the wood seems to balance with the hardness of the stone.

These days, it seems easier and quicker to lay a new prefinished floor over what you already have. But a good well-finished wide-boarded floor has a certain feeling of "I'm here for eternity." It gives soul to a space and will warm up many a sterile stark interior. The white-painted floorboard gives a feeling of a country or seaside interior. The black-painted floorboard gives a slick, more urban feel and creates a hard-edged crisp living space.

In open-plan spaces, which are being used more and more these days, the use of different flooring materials on the same level is a way of defining the space for certain uses. A wood surface can be used in the living and sleeping areas, and a tile or stone surface in the kitchen and wet areas; common sense dictates the appropriate product.

With the conversion of more and more industrial buildings, the reuse of existing flooring is well accepted. The old flagstones of a barn, the polished or painted concrete floor, or poured rubber floor, usually found in operating rooms, are now becoming common options for the home.

The material, the color, the price, and the availability all add to the effectiveness of a new floor. If you can lay or install a new floor yourself, it can be a pretty immediate project. If you want something a little more special, calling in the experts can be the best way. A terrazzo floor covering a complete space gives a cool and calm united space. While it is a lengthy and costly project, it is well worth the time and money.

It is amazing how so much good flooring has been covered up over the years. Whether it be wide old oak boards, 1930s herringbone pine parquet, or Victorian encaustic tile, it is entirely up to you to restore and keep what you have,

8 Brick flooring laid in an alternate grid fashion gives a sense of space and proportion to this kitchen floor. **9** In a former industrial building the shabby concrete floors were updated with a coat of floor paint. The color was matched to newly laid concrete to resemble the original floor.

10

11

12

10 Limestone squares run throughout the entire ground level of this house. They act as a great background for the mid-20th-century modern furniture. **11** Clear or pigmented epoxy binders can be fused with multicolored or natural stone; this one is marble and resin. Many such floors are fast-curing and unaffected by sunlight, which makes them popular for outdoor use, too. **12** Poured terrazzo floors used from around 1930 to 1960 are often found under newer floor coverings. Luckily, the material can be easily restored, re-polished, and filled if necessary. Terrazzo is often used for countertops and is now becoming a popular alternative for a new floor.

13 Here the use of a natural Berber carpet complements the stark brick feature of the chimney rising from the floor below. It also balances with the grid of the Bertoia chair. **14** On a concrete floor a new runner has been placed. Besides making the floor kinder to the feet, it leads the eye into the gallery area of this apartment. **15** The color of this carpet was selected to tie in with the landscape, to bring the outside in. **16** This new carpet is silk with metal thread detail, a fresh take on a traditional carpet method, bringing it up to date. It also introduces some pattern to an otherwise unadorned interior.

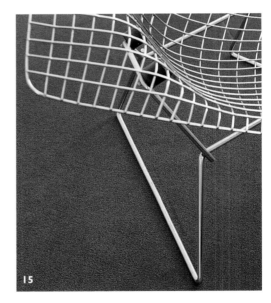

to replace it with a newer version, or to completely change the floor surface. After all, the floor of your home takes up a vast surface area of your space. It has to be comfortable underfoot and pleasant to the eye, and work in the area where it is allocated, this is particularly relevant in areas of the home that incur a lot of traffic. The use of flooring to define the space within is a good and less disruptive way of making a home more modern.

STAIRS

I In this Tribeca loft in New York, the staircase follows the diagonal partition wall with the neighboring unit. It cuts through the original roof line and enters into the new room. **2** The steel treads have been drilled with circular recesses to act as anti-slip devices. Here you can see the simple steel attached to the wall. **3** When you enter the loft, the staircase leads your eye to the upper level, but with the void below you also note that it descends to the lower quarters. The mix of steel for the upper treads and wood for the lower relates to the construction of the original building.

In years past, downstairs was where your staff of servants carried out their duties. Downstairs today could mean storage, home office, utility room, or even a downstairs bathroom. The staircase now is not just a facility for getting from one floor to another—it can also be a decorative item of visual effect for the modern home, and can act as a valuable area of dual use. It will take you up and down the stairs, but can also be your office space.

As a means of getting from one level of the home to another, the staircase can be straight, spiral, in a recess, in the center of the room, or hidden in an enclosed hall. The proportions must be correct, and the safety aspects must be considered. There are certain constraints regarding the riser and tread dimensions that may need to be taken into account. But when it comes to the visual importance of the staircase, it is the materials that can give it visual impact.

Nowadays, the staircase is a feature that many people will consider altering, and whether to relocate it entirely or just to re-finish it are just two of the options. With open-plan types of accommodation being increasingly popular, the staircase is a more visual element and has a more important role in the modernization of the home.

4 Rising up against a rustic stone wall, a simple modern staircase made from steel 'I'-profile beams has had trimmings added for the treads. The handrail was applied in situ. 5 The typical Victorian staircase has been updated here with colored paint and polished treads. The ornate bannisters give the interior a sense of the past.

The use of glass as a contemporary structural element as well as a visual component has helped to increase its use for staircase construction. With glass treads, balustrades, and landings, a glass staircase can look stunning and bring much-needed light to an otherwise dark hallway.

The use of steel as a staircase frame has also helped create the modern staircase. With treads of metal, plywood, or any type of wood, a metal-framed spiral stairway can look timeless in a classic setting. With many alternatives available off the shelf, the metal staircase offers an almost instant transformation. With the help of a welder, any shape or proportion of construction can be created, making it unique to the space in which it is installed.

When it comes to the banister and balustrades, you have the choice to either enclose the staircase or leave it open. A solid side can give valuable wall space and also create storage space beneath. The open tread gives a feel of openness, especially if the staircase is within an open space.

In my experience, the relocation of a staircase can make the most drastic modernization to any home. In an older house the basic staircase can take up so much room that to re-configure it and maybe move it can open up the space in a whole new way. Turning a staircase can give a larger, more welcome entrance hall with less need of a turn or a half landing, and the modern elements in its construction inevitably make the whole place feel more modern.

6 A spiral staircase rises up from the main living room to the upper bedroom level. A spiral, while taking up less floor space than a traditional staircase, can dominate a space, but it does allow the eye to see through it. **7** This built-in staircase follows the line of the circular room. Its monolithic construction makes it both a visual feature and a structural one. The treads are made of slate. **8** This staircase divides into two, one side going to the bedroom area and the other leading to the outside terrace. Constructed from simple off-the-shelf elements, it has been customized with a solid bannister on one side and tension wires on the other. The open side allows light to penetrate through to the level below. **9** An oak staircase made from simple slabs without any molding or detail appears to float between the two white-painted walls. There is a gap on each side that accentuates this illusion. A tubular stainless-steel handrail helps you climb the three-story staircase.

2 Overlooking San Francisco Bay, the windows in this building are the walls. Completely glazed elevations are broken up into a grid of opening and fixed planes. The overhang of the roof acts as the all-important sun shade.

I In this Los Angeles home, a mid-century classic, the full-height sliding doors open up the indoor room to the outside, creating that important indoor/outdoor lifestyle.

Natural light is a valuable element of the workings of the space within the modern home. The style of the windows in the home are inevitably inherited with the building you buy. In a house, the front elevation is usually within a row of other houses, and in my opinion there is nothing worse than one house on the street that has had its elevation altered with poorly proportioned windows. It is always best to keep the originals on the public face of a building if you can.

The back elevation is a different story, though. Here, you could change every opening to create larger windows, to create a different format of windows, or to make the whole back of the house into one big window. Getting the balance right is the clever part. The best way to do it is to consider the windows if and when you add on an extension, to tie the old in with the new. This would be a natural progression to modernization. With so many types of construction and materials available, you can get a good match to the originals or a good

3 High-level strip windows are great for areas in which you want privacy, especially when you live in a built-up area. They bring light in without the worry of being overlooked. In this house in Sarasota, Florida, the clerestory window runs the length of the living area along the street side of the house. **4** An original feature such as this six-sided window gives character and soul to this renovated space. Retaining elements such as this keeps you in touch with the origins of the building. **5** This curved wall of window features custom-built bent sections of glass. Built in the 1930s, this bay was included to capture the distant views of the sea beyond and the surrounding landscape. The window ledge is poured terrazzo. **6** In what was once the shed of a 1930s house in Frinton-on-Sea, Essex, an oval curved window reflects the curves of the building, as does the circular skylight.

complementary style. In some ways it is best to stick to the same materials, but perhaps to change the format. There is a huge variety in the style of window that has been available throughout recent history, from the wooden sashes of the 19th century, to metal frames of the 1930s and 1950s, and vinyl windows of the late 20th century. Even the UPVC windows being widely used these days can look great if used in the right place, and with the right proportions within the elevations.

7 In an addition to a Victorian school building, this modern wall of glass is softened with Venetian blinds. This area opens onto a courtyard garden. **8** Light from this hallway is let into an internal room via this high-level window strip. **9** In a room that is overlooked, a small grid pattern helps distort images. The blind provides total privacy. The grid of windows is an original feature of this house. **10** This full-height panel of glass allows necessary light through to an otherwise dark stairwell.

DOORS

I The original front door of this San Francisco townhouse provides an immediate understanding of the house you are about to enter. It completes the Victorian façade, and the modern pieces placed nearby show how effective mixing styles can be. **2** Pivoting doors provide a wider opening. This entrance door has a woven steel insert. In this lobby, the front wall is entirely glazed with various openings. The lower-level window acts as a ventilator, capturing the cross breeze. **3** In the same house as above, the walkway to the upper levels has a wall of windows that have louvered screens, which work as sunshades as well as privacy panels. This gives a serene atmosphere to the whole space.

When it comes to doors,

the use of an original-style door on the front elevation is always a good start. But when you get inside, the internal doors can be whatever you want. You can have conventional hinged doors, sliding doors, folding ones, or those that roll into a pocket at one side of the opening. Door frames can be removed or adjusted to give floor-to-ceiling openings or relocated to open different aspects of the rooms in question. They may even be used to unite or divide two or more areas.

The materials are so varied that there is no reason for using the wrong door. Glass or glazed internal doors will bring essential light to dark areas and gloomy corridors, and to create a more naturally flowing link between each room.

In a more open-plan arrangement, sliding doors can give the appearance of a lot more space.

4 In a Tim Seibert house built in the 1960s in Sarasota, Florida, this area within the house has sliding wood doors that mean this space can be either enclosed for privacy or opened up for entertaining. These panels slide into pockets so they can be completely concealed. **5** The top floor houses the bedrooms and main bathroom. The master bedroom spreads across the back of the house. The use of a sliding panel as a door unites the spaces when open without taking up valuable room. When it is closed, it looks like a seamless part of the wall. **6** This Amsterdam residence has many sliding doors and panels, facilitating a totally flexible space that can be adapted for many uses. These deep red panels run along the second-floor landing.

7 These aluminum-framed glass doors run along a track carved into the stone floor. Keeping a uniformity throughout the areas on both sides of the track also retains an uninterrupted floor when the doors are opened. **8** Three glazed panels make up the entire screen wall to this area. The dining area can be completely exposed or cut off from the kitchen, ideal for entertaining when a more intimate atmosphere is required. A single fixed glass panel runs along the staircase area.

A pocket door is even more space conscious—the wall taking the door needs to be thicker, to provide enough of a recess, but the actual wall on each side of the door will remain clear for other uses.

The door is an important part of the modern interior. While to some it is a great idea to have no doors, just openings, the flexibility of being able to open up and close off a room or area does make the overall space feel larger.

On the whole, the windows and the doors of your modern space should complement the architecture and materials of the original building, but they must also do the job required of them—to bring in light, to ventilate where necessary, and to divide or unite the areas within the home.

WALLS

For me it's the walls within the space that make it a home. I've been in many homes where the owners have instantly removed every wall so that they are living in a barnlike space with the furniture pushed against the exterior walls and pointless groupings dotted around. With open-plan living so often being one step forward towards modernization, the balance of which walls are removed, which walls remain, and where walls are installed is so important.

In my loft space, I am fortunate enough to have a very thick structural wall that runs the length of, and divides, the entire space. This gives a natural break to the space. Compared with my neighbor's unit, which has no division, it feels twice the size. This is because my neighbor's loft is completely open—there are no hidden or surprise areas. What you see is what you get.

The placing of walls is very important. Whether they are solid, sliding, floor to ceiling, or at low level, they are there to define the space and the use of the area within. The sliding wall, of course, creates the most flexible space. To be able to open up or enclose particular areas of a home with moveable screens, either solid or opaque, keeps the entire space usable. A space with no dividing walls gives less scope to place furniture, artwork, and entertainment equipment.

1 In an industrial building, now a home, the original brick walls have been painted to bring them up to date. Layer upon layer of paint has been applied to achieve a uniform even finish. **2** The concrete block wall, forming a large-scale grid pattern, makes a great backdrop to contemporary furniture and furnishings. Here it has been painted dark red to create a dramatic interior.

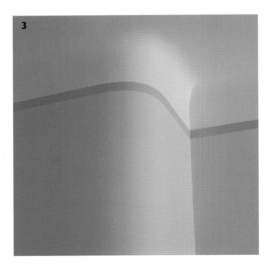

3 Paint techniques have been used to create a false molding. The wall blends into the ceiling and the highlighted line gives depth between the planes.

4 The fireplace wall divides this house, with the kitchen on one side and the living area on the other. The use of color exaggerates this division. **5** The subtle finish to this wall unites the painting and the staircase. It is pale grey with a pearlised effect. **6** Retaining the 1970s bronze-finish ceramic tiles works well in this contemporary bathroom. They complement the heavy timber beams reflected in the mirror inserted into the grid at the appropriate height above the sink.

When taking on an open space, it can be quite difficult to decide what to do. The best way to start is to break down the space according to how you want to use each area. Also decide how flexible you want each space to be and work out whether the walls should be solid, glass, moveable or fixed—these decisions are all part of the total concept. A series of walls can create visual space, and make your everyday functions easier from the start.

The treatments used on walls can make them stand out as features or disappear so that they are simply a division of space. Deciding what to use on the wall surface is always a difficult task. If you paint, what color, what finish, where to stop and start

7

8

7 In this Finnish apartment by Ulla Koskinen an opaque glass wall allows, in the daytime, light from the living room to flow into the bathroom, while it acts as a full-length illumination at night. **8** These industrial pre-formed glass panels are set in a curve in this extension. They give a distorted view, which diffuses the outside, whatever the weather. **9** On the top floor of a San Francisco townhouse a bathroom with a glass roof is concealed behind this wall of glass. The midday sun is so strong that it almost appears to be flood lit.

10 In this bathroom, light floods through the full-height opaque glass wall. **11** With all traces of plaster removed, the old brick walls in this Victorian house add character, texture, and pattern. **12** A selection of colored tiles laid in a random pattern gives a great sense of disorder to this well-planned bathroom. Tiled floor to ceiling, it is both practical and unusually exciting. **13** The concrete block wall in this Gene Leedy-designed home runs from the outside in. A brick floor has been laid in the same fashion to complement it.

the color? If you wallpaper: what style, where to use it? If you panel, what should you use: wood, textured panels, glass, Plexiglas, or even metal? Any application to a wall must be done for the right reason.

The use of color on the right wall is a task in itself. To paint every wall white can look great but it is an easy option. The highlighting of different adjacent walls can add further dimensions and give illusions of space in many interiors. Oliver Hill, an architect and designer working in the 1930s, would use various pastel colors on walls that would balance and complement the light coming into the room.

Experience the spaces before you apply any decoration. You can see where the light falls and reflects so you can try out colors to get the best results. Even the use of various tones of one color can add dimension and space to a room.

FOCAL POINTS

A room with a view does make sense, even if the view is contained within the room. The focal point—a fireplace, for example—gives a focus to the room. It is somewhere to start. Okay, not every room has a fireplace, but in the main living area it often does exist, so the fireplace is an obvious focus. What should you do to make it more modern and update your interior?

It's good to have this type of focal point, but a traditional Victorian mantel can be somewhat sad in a modern interior. A good simple marble fireplace is timeless and would give a classic modern feel to any room. With the use of

1 This wide fireplace of steel and stone is a heavy welcoming sight when you enter this room. 2 A simple wood-burning stove has been concealed behind a stainless-steel sheath to enhance its appearance and to create a focal point in the open-plan space it which it stands. 3 This massive chimney breast dominates this space, yet it works as a visual screen to the entrance area of this house, affording privacy when necessary.

3

different fuels the fireplace can now be easily installed even without a chimney. The hole-in-the-wall style fireplace has become popular recently. Referring to the 1970s for its look, the high level slot can easily be used for other purposes besides heating. And with more ready-made surrounds now available there is one for most styles and budgets.

Today we often see the plasma screen as a focal point in the room. This is so much better than the television set that, until recently, was often the central attraction in the main living room. With the use of this type of entertainment equipment, viewing at least it is quite discreet, and is really now treated as a decorative element to the interior.

Bookcases or display units can also be focal points. An alcove can house an instant installation especially created for your collection of historic pots to form a focal experience. A focal point could also be a great painting, or a sculpture on a plinth, for example. This is what draws the eye into the room, and gives an ambience to the whole space.

5

4 The plasma screen (pictured opposite), the new focal point in today's home, has been set into this three-quarters-high wall. The home now has a new necessity. **5** The long, low recessed shelf functions as a focus in the same way a fireplace would. It is also good for display, and it would give an instant atmosphere to any space.

LIGHTING

1 On the landing and above the staircase of this modernist house, rows of low-voltage spotlights have been installed parallel to and close to the wall edge, thus bringing light to the rising staircase and the artwork. **2** Recessed group lighting that is flush with the ceiling creates a neat alternative to the individual downlighter.

The importance of lighting cannot be

overstated. There is nothing worse than a dimly lit room with inadequate lighting facilities. Whether you are using uplit, downlit, concealed, central, floor, wall, or occasional lighting, or ideally a clever combination of some or all of these, you can light up your interior any way you want. If you are doing major refurbishment work on your home, the introduction of downlighters or built-in lighting is a good option as all the work of recessing unsightly wires can be done while the ceilings are down and the floors are up.

The most important principle of lighting is the use of the right type of illumination for the task in question. There are basically two types of lighting: directional and purely decorative. It then breaks down further into recessed, surface mounted, concealed, or freestanding styles.

With the recessed method there are endless options for low-voltage, low-energy, or halogen energy sources. For surface lighting you could use can downlighters, track-type spotlights, or low-voltage lamps suspended on cables. This category also includes decorative center and wall lights.

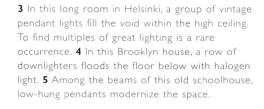

3 In this long room in Helsinki, a group of vintage pendant lights fill the void within the high ceiling. To find multiples of great lighting is a rare occurrence. **4** In this Brooklyn house, a row of downlighters floods the floor below with halogen light. **5** Among the beams of this old schoolhouse, low-hung pendants modernize the space.

3

4

5

Concealed lighting would give overall lighting in display areas, on staircases, and at strategic areas in the home, such as at different floor levels where the use of florescent lighting could be a good idea. Freestanding lighting includes table and floor lights. This type of lighting offers the opportunity to give atmosphere to the room. Whether you choose contemporary or vintage examples is your choice.

Throughout the 20th century, there have been key examples of iconic lighting. In the 1930s the British Bestlite was featured in all the best interior and architecture magazines of the time. Available in a desktop, wall and floor-standing version, it covered all needs. In the 1950s the

6 In a contemporary dining room in a Victorian row house, a mix of modern furniture and period lighting creates an individual style. **7** Two types of lighting illuminate this small home office. The uplighter will flood the room with bright light while the contemporary crystal pendant will create an intimate atmosphere. **8** Suspended in mid air, this zeppelin-type pendant floats beneath the Victorian rose in the entrance of this San Francisco townhouse. **9** With all this daylight a great deal of electric light is needed to compensate when the sun goes down. Here six pendants are suspended at random heights from the beamed ceiling.

George Nelson bubble light hung from the best ceilings throughout the United States. In the 1970s, the Arco light arched over various arrangements around the world. Now with so many classic modern designs being reproduced and re-introduced to the market, the options are vast when it comes down to what light to choose, and where to display it.

With directional lighting, the options are also endless. With a vast range available, and an increasing number of sources, from the local hardware store to the architectural lighting specialist, there is something for every budget. Whether you opt for the low-voltage, angled downlighter, floodlight or the recessed spotlight, it is up to you to select the correct installation for your space.

10 Angled to mirror the beams, this Italian production task lamp is ideal for bedtime reading as it swivels and tilts wherever you may need it.
11 This grouping of architectural paper lamps by Noguchi fills a corner of this Brooklyn townhouse. As well as bringing light when needed they bring a timeless atmosphere to this family room.

DISPLAY

If you have something to show, why not have it on display? To create an area purely for display is an instant way to make your interior an individual one and inject personality. Whatever you may collect, whether it be books, pots, African art, 1970s glass, or 1980s Barbie dolls, having them on display can enhance the atmosphere of the room or space.

Getting the balance right is the clever part. A whole room dominated by glassware can be too much, but one alcove of it can be a great backdrop to your other treasures. Deciding how and where to make your display area can be the tricky part. Of course you want the arrangement to be in a place where you are going to appreciate it, but is also needs to be located in an area that is not better utilized for something else, more useful – for the television and sound system, for example. How you should house your display is another challenge to be solved. You can buy various units of furniture specifically designed for this,

2

1

1 On this rosewood chest by George Nelson is a well-grouped selection of family heirlooms and flea-market finds. On the wall hangs a new work by Jo Shane. **2** These items have just happened to arrive here. Containing a selection of items within a tray or frame creates an instant display. **3** Behind the sinuous deSede sofa, a shelf runs the length of the room, allowing a constant change of artworks. Here a mix of photography and other media are on view.

4 In the lobby area between the living and bedroom spaces in this house in Sarasota, Florida, a group of ethnic and tribal art provides a dramatic focal point. The massive shelf is built into the alcove walls.

5 In the kitchen, display is often forgotten. Decorative vessels look good on display even without the fruit they might sometimes hold. The items in this kitchen all carry a pierced element as a linking theme. **6** Chain-store shelves are fixed up above a bed at different spacings. This arrangement enables many permutations of display, including paintings, pottery and ethnic basketware.

whether it be a series of shelves for an alcove, or a complete freestanding unit with individual sections to divide, and categorize the collection.

I always feel that a mixed selection of items looks best. Include books, some art objects and artefacts—maybe even integrate the various parts of your hi-fi system as well, to make it look less obvious, and as though your collection has grown and developed naturally over the years. Having a mixture of items makes the display less precious, and more approachable for your guests and visitors to appreciate.

Other elements will have to be taken into consideration. Lighting, for example, is an important part of a good display area—whether you use spotlights or illuminate with concealed lighting depends on your own taste. Directional spots give you more flexibility as you can angle the lamp on

your favorite pieces. With concealed lighting, you can usually back light the whole arrangement, which in itself can visually become a focal point to the room.

In some cases, a whole room can be called a display, especially if you collect classic furniture. A group of key pieces of 1950s furniture in the right setting is an instant collection on display. It may be a bit precious, but in an open-plan loft space this can work very well, as there will be enough space for useful seating arrangements.

It all depends on your level of obsession. Some of us are forever looking for the next object to add to the collection. So in those cases, a flexible system would be more suitable. An area that perhaps has open and concealed areas is maybe best suited for this, so that you can have your favorite treasure on display, and conceal the rest behind closed doors.

STORAGE

1

1 This unique idea, in a Victorian house in San Francisco, works as book storage and also a balustrade to the staircase that sweeps down behind it. The units are suspended between a light steel structure. **2** In this white room, storage on wheels is a flexible solution. You can wheel it to the utility room when you run out of clean clothes. **3** In a house in Sarasota, Florida, which has been extended by Guy Peterson, an entire wall is given over to cubbyholes for the gathering of favorite books and objects. This area is not an arranged display; it is a forever changing full-height backdrop to the room.

With the clutter-free, minimalist interior inspiring most of

our choices in our home today, storage has to be the most important element to make the modern lifestyle function effectively. Think about where you are going to need storage, what it is to be used for, and how often you will need to access it. The choice of either open or closed storage, and for built-in or freestanding storage units all depends on the surroundings in which they will sit.

Storage problems are solved so simply. With off-the-shelf systems so easily available from most of the chain stores, you can buy a complete wall of storage and take it away with you that day. Of course, you can get the carpenters in to do this and that, but why bother when you can get the ideal system so reasonably? Using existing alcoves for storage seems like an obvious solution,

4 In the small hallway in this Helsinki apartment, full-height cupboards have been installed. Although they are quite shallow, they can still contain many items. **5** In the entrance lobby of a loft space in New York, this complete wall of veneered panels disguises both deep walk-in storage and regular racked-out closets. **6** Understair storage is often hard to access. With both pull-out and open-up storage, here every square inch is usable.

but to use a complete wall of storage units as a divide between two existing areas is an advanced idea and utilizes your space to the full. Those of us who live in a reclaimed industrial space without the usual domestic framework appreciate the importance of good storage.

The items you are storing will dictate the type, size, and scale of storage you are installing. You could have concealed hardware to doors so that the area looks like a flat wall. You could have a grid of small doors to make the whole wall of openings look more decorative. The doors could be wood fronted or veneer-faced board; they could be painted to match or to contrast, or they could be glass.

Open storage can look good when combined with groupings of display items. Using freestanding storage units can work well if they are the correct proportion and style for the room. Using vintage pieces especially can give a great feel to the room as well as being useful. A long sideboard will provide both storage within the cupboards and display space on top. One of the many combined storage systems

7 In this house in Florida, the plasma screen is mounted above a custom-built metal-paneled unit that pulls out, and opens up, and houses all the entertainment equipment. To add to its architectural feel, the top of the unit is fashioned out of concrete. **8** The entertainment in this London house is instantly available, or it can be hidden when not needed. It is concealed by walnut-faced sliders. **9** A long, low sideboard is available for storage, but also doubles up as a great surface for displaying favorite treasures. A pair of identical mirrors hangs above. **10** To use a partition as storage is an ingenious way to break up a space; behind this wall of closets in the bathroom is the master bedroom.

dating from the mid-20th century can solve a great deal of problems, with a unit that includes open shelving, drawers, drop-flap cupboards, and even a desk area.

When you look at room with storage in mind, first always try to figure out your furniture layout. You may deprive yourself of valuable wall space if you have pieces that need to stand against a wall, or if you have favorite artwork you want to hang in a prominent place. Organizing your lifestyle, and clearing out all unwanted items is the way forward. Storage should be discreet, and exploit areas that are not good for other uses. Use the space under the bed, and disguise the ugly boxed-in areas for the boiler by encasing it within a larger storage unit. You could use the eaves of the roof for a wall of recessed storage.

FURNITURE

One of the easiest and quickest ways to make your home more modern is to update your furniture and furnishings. This could be an instant transformation, such as updating your old sofa with a new cover in a great color or print, or putting up new window coverings.

In a furnished rental situation some of the furniture inherited with the lease can be pretty basic. But with a bit of time and money, that old seating arrangement could be transformed into the ideal relaxation area with a few modern pillows and a textured throw—how easy is that?

In a more traditional home, perhaps a Victorian row house, the placing of good-looking items of furniture can be a great balance to the original features of an older home. A really good piece of 1920s design—the Barcelona chair by Mies van der Rohe, for example—can gel surprisingly naturally against a molded and paneled fire surround.

2

1 Furniture, as well as being comfortable, has to look good. The Arne Jacobsen egg chair, here covered in hide, gives the added curves to this otherwise linear interior. **2** Custom-made furniture does in most cases solve the problem of finding the right piece for the right place and making sure it does the job required. This low-level deck for the plasma screen does not interfere with the window feature in this room.

3 On a deep shag-pile rug, the classic Noguchi coffee table makes a sculptural statement. **4** You do not need the originals to create the right interior. Here, in a mid-fifties house, reissues of furniture by George Nelson, particularly the day bed, have been used. The amoebic table softens the hard corners. **5** A see-through chair by Bertoia needs a pillow for added comfort, while its graphic qualities are inviting.

Choosing furniture can be a lengthy and tedious task. Before you buy it is better to try to experience the space for a while until you have worked out how you use it and made a list of your requirements. The size of the sofa, the type of dining table, the colour of the rug and the window treatments: these are all the sorts of decision you will need to come to once you have assessed your needs.

In your space it is the selection of the furniture and furnishings that make it your home. Throughout the world the desire for modern styling has been so great that many classic pieces of modern design are so widely reproduced, some by the original manufacturers and some by the firms that do various versions of the original. If you simply want to create a look, the non-original versions are ideal,

6 The Le Corbusier Basculant chair always looks good, wherever you put it. Designed in the 1920s, it is hard to believe it still looks so modern. In this library, its framework relates to the uprights of the book shelves.

7 Poul Kjaerholm chairs are classics from the mid-1950s. Their streamlined profile works well in any material. These chairs also come in leather or canvas. The table is by Mies van der Rohe.

8 In this classically proportioned room the Bull chair by Arne Jacobsen hovers in the wings. In the room there is an eclectic mix of styles, creating an individual style, but every piece relates to the next. **9** The plastic stacking storage system by Anna Castelli has recently been reissued, making a visual and useful bedside cabinet. The chair keeps the circular theme going.

as they do look good. But they do lose their value as soon as you buy them. If you are going to buy classic pieces of furniture as an investment, make sure you buy them direct from the original manufacturer. These pieces will certainly retain, or increase, their value.

There is so much good new design around now that whatever you are looking for there is a wide range of price and quality to choose from. From Scandinavian chairs to the American original manufacturers, you can select from a vast array of good design. A mix of both is a clever way to do it. Get your key elements of storage and display from the chain stores and select a few iconic pieces from the top manufacturers.

New design around these days relates so much to original pieces that perhaps if you do not have the time to select good vintage pieces, non-original modern examples are the best option for you. There are many sources, but maybe working with a good furniture and design dealer is the way of getting what you really want.

I believe it is always a good thing to try to find less obvious pieces of design, perhaps from the lesser known, but still great, designers. The way forward is to include key modern pieces in your interior, whether original or reproduction. On the whole its the look that counts.

Of course, condition also matters. In a minimal, clutter-free environment, the last thing you want is a worn-out piece of furniture, but in an ecclectic, ex-industrial space it would look fine.

Together with good design, good condition and with the furniture well placed within the space, the modernised home will be complete.

8

ZONES FOR MODERN LIVING

Page 152 The curved kitchen design makes a dramatic visual statement while offering a seamless worktop within a self-contained space. **1** The sleeping zone needs to be inviting, to have ample storage and have good lighting for bedtime reading. Here a wall of storage acts as a head board and disguises the flue for the boiler below. **2** The kitchen in many homes now is also the place we eat. This kitchen has space away from the wall of units for a small table and chairs. **3** Relaxing at home can often be a difficult thing to do. To find the time and even more to find the comfortable seating to relax in can be a hard task. In this Helsinki apartment by Kari Lappalainen massive seating units appear to fill the space, but they do look inviting. **4** In the bathroom, really, anything goes. As long as it has somewhere to do what you do, there are many surfaces and fittings you can choose from. Here, economical porcelain fittings are put together with iconic taps by Arne Jacobsen. The bath wall is faced with solid douglas fir while the others are just painted.

The zones for modern living reflect the activities we carry out in our homes, and to which in traditional homes completely separate rooms are devoted – they are obviously for relaxing, cooking and eating, sleeping, bathing and working. Which ones you develop is up to you. If you are working on a space new to you and you are living in it at the same time, you have to decide on your priorities. If you work from home you may need to get that space established first and foremost so that you can get back to work as soon as you can. If you can work elsewhere you may want to get the basics, such as plumbing and electrical works, done first.

It is perhaps more commonplace to get all the services in place as soon as you can, but sometimes it is hard to make decisions without having first experienced the space properly If you are looking to modernise your existing home, at least you have had time to think about what you really need from the space. Is the bathroom in the right place? Does the kitchen layout work for you? Is there enough storage in the relaxing area? Do you want an ensuite shower room? Will you need more drastic alterations such as stripping out or moving walls? It all really takes time to work out.

Modernisation through changing or updating decorative features produces less disruption and as you are not doing any structural work many zones can be worked on at once. Getting a new floor laid throughout the zones, for example, will unite them and instantly transform them.

The choice of which zones to develop and when is a personal one. To some a great relaxing zone is more important than the cooking zone. If you have no desire or urge to be a master chef why waste time and money creating a super kitchen if you are not going to use it? If you do not work from home.just make sure you have some space for the home computer. Of course the sleeping and bathing zones need to work in any home, but there are simple and cost-effective ways to work on these areas. In both spaces storage is always important. Somewhere to put things can only add to the benefits of modernising your home.

Breaking the home up into zones makes it easier to work on one area at a time, even though you do not want to treat them as individual spaces. You want one to flow to the next. You want them to be, on the whole, flexible and you want them to enhance your everyday modern lifestyle.

RELAXING

One of the most valued assets of having your own space is to be able to create the most relaxing areas you can within your own home. To be able to come home from work, or to have finished working at home and moving into or settling into the relaxing zone is the most important part of living in, and enjoying, a modern home.

This home must do all you want it to do. But creating a zone to chill out in is perhaps not a simple task. There are many aspects of the relaxing zone to consider, such as the colors and patterns to be used, the type of decoration, and even the introduction of built-in furniture.

Within the home the relaxing zone really needs the most thought since it will be used for so many different activities. If you are just doing a quick make-over of this space—purely decorative modernization—all you are really going to consider is the color and decoration, the flooring and other surface finishes, and perhaps new or updated furniture to work within the improved space.

The decoration in the relaxing zone needs to be cool and calm, but perhaps with a focal element of pattern or other color to give it personality. You do not really want it to be too stark and organized as this is the zone where you will be entertaining most probably, where you will be reading the Sunday papers, and perhaps where the children will play. A little bit of visual clutter can sometimes make the room. You will need storage areas, display areas, space for the television and sound systems, and, of course, comfortable seating.

If you are going to start to update the features of the home, you may want to remove an existing fireplace so the television screen can go in its place, you may want to remove a dividing wall to give you more space to relax in, or you may

I In this In this mid-1950s Florida home, walls of glass let the outside in, and increase the living space. Outdoor furniture also becomes as important as that selected for the interior.

2 The raised level of a former tram shed in Amsterdam makes an ideal place for an area to relax in. Away from the service areas of the space, the grouping looks appealing. **3** The L-shaped house allows for a second, more intimate sitting area. With a wall of entertainment equipment this area is ideal for the kids or the grown-ups. The low-level table and all those pillows encourage the use of the rug for extra seating. **4** In the roof-top room in this New York loft a glamorous sectional sofa zig-zags around the room. Facing the windows, it gives a great outlook onto the New York skyline.

you may want to install an entire wall of dedicated storage to house all the unsightly entertainment equipment.

Re-organising this space could give you grander solutions. Besides removing a wall you might want to totally configure the layout of the main floor of your home so as to create a more open-plan space for relaxing in. To make a flowing plan from one space to another, and perhaps linking this space to the cooking and eating zone and maybe even the working zone will create an ideal environment for 21st-century living.

The ultimate in making a relaxing zone stand out from the rest would be if you were to extend your space to create a great light-filled room, that went out into the garden space, that included an outdoor sitting and relaxing zone and that naturally continued the flow of the house outside.

5 The bay window in this room is in a good position for the sofa. This wide one houses a seating arrangement that has moveable elements so each user can create their ideal sitting position. The window can be closed off with Roman shades to create a cozier atmosphere. **6** This space is one total chill-out zone. With the mix of shapes, color and graphics, the entire interior is visually stunning. The long low seating element allows for group entertaining.

5

The important things to consider if you are building an extension to your home for relaxing is that it be accessible, that it has a purpose, and that it is inviting to both the eye and the body and, of course, the inevitable comfort it can give.

Whatever way you decide to modernize your home, the furniture and furnishings you select for this zone must work. Whether you update what you have with new coverings and finishes, install new window treatments, or just add bold cushions or a wacky carpet, these are all important elements for this space.

Relaxing, comfortable seating does seem to be the most difficult item to find. There are many great-looking sofas available, but are they comfortable? Seating units are at the top of the list when selecting furniture for the relaxing zone. Do you want vintage or contemporary, sectional or conventional? It is up to you. If you have a good comfortable seating arrangement already, perhaps the best thing to do is to have it reupholstered in a new color or more modern fabric. If you are buying new, test drive as much as you can. Maybe a friend has a great-looking sofa? Try it out. But even if you are stuck with your old faithful sofa, a new batch of patterned throw pillows can add that bit of extra visual and physical comfort you have been looking for.

The relaxing zone might also include your eating or working place. These areas have to gel with the main space, and not interfere with the overall effect. By using similar or complementary styles, fabrics, and colors you can create a space of harmony and relaxation. The relaxing zone must be seriously considered: it has to be inviting, accommodating, and work for you, your friends, and your family.

The eight hours of sleep we require a night need to be experienced in the appropriate setting. Whether you are creating bedtime quarters for yourself, your children, or your occasional guests, the sleeping zone is much more than simply a bed placed in a room.

When you are deciding how to modernize your home, the sleeping areas are perhaps the ones to consider first. After all, getting a good night's sleep will get you through the next stage of development and also give you time to mull over decisions. Not that you need to have sleepless nights while working on your home, but I find it is often at night when great ideas occur. I always keep a pad and pencil next to my bed to jot things down or sketch a layout or fixture detail.

So the space could be for yourself, your children, or maybe an office-guest room. These days, with our flexible ways of living, we always seem to need to incorporate so many other things into our rooms. In the bedroom, beside the bed we need ample storage for clothes, adequate areas for television and entertainment equipment, a mini library, good lighting for nighttime reading, perhaps a small work area, an area for fitness equipment if you are so inclined, and then, maybe, if there is enough space, a connecting bathroom.

This is just in your room. If you are working on a child's room, you will need much more storage for toys as well as clothes, more flexible sleeping arrangements for a sleepover,

1 This bedroom opens onto a wraparound terrace that overlooks San Francisco Bay. To wake up and experience this view must be exceptional as it is ever changing through the seasons and the times of the day and night. 2 In the center of this space is the bed. This sleeping zone is also extra seating when entertaining and is used by the owners to watch their favorite television shows This arrangement calls for strict organization.

2

3 The windows of this bedroom resemble artworks hanging on the wall. With the tropical outlook you know where you are. The built-in bed and side units keep the room well organized and easy to maintain. **4** In the center of this small house, one of the bedrooms has high-level glass panels to let light in. Used as an occasional bedroom, it is divided from the next room by a small shower room. Note the flush ceiling speakers for total ambience.

and a suitable area to do homework. In a guest room, a sofa bed or one that folds into the wall keeps the space more useful for other activities such as the home office.

All of these are key elements of the installation of a sleeping zone. The decoration of these areas is also as important. As in the relaxing zone, color and pattern are important. In this zone it all depends how you like to sleep— you are, after all, creating this space for yourself. Some like to be in total darkness while others like to be awoken by bright white morning light. This, of course, can be controlled by appropriate window treatments, but on the whole, when keeping or making your home modern, a brighter decor is the ideal solution. You can darken it when you want to sleep and brighten it up when you need to wake.

Going back to basics, the bed of course needs to be right for you, your sleeping patterns, and your personal preferences as to a hard or soft mattress. It also depends on the size of the space you have designated for this zone. With such a wide range of sleeping products available, there is something for everyone and every budget—whether you

choose a twin, queen, or king-size bed, it's all up to you. It is what else you introduce into this zone that will make this space modern and personal to you.

If you are looking for the quickest and simplest way to modernize, the use of great fabrics and paint can make your zone comprehensively more modern. The decoration is what you will initially see, and it will create the atmosphere and ambience of the room. To create extra storage, you could use the space under the bed. You can mount a television on a wall-hung mechanism, you can add new bedcovers and window treatments, and you can install new lights. All of these are simple projects that are quick to complete.

Updating the more permanent features of the sleeping zone would include replacing old-fashioned doors on closets and storage with new streamlined or even mirrored alternatives, installing a new ambient lighting system, or even removing a fireplace to use this space for an entertainment storage area. Of course, whatever way you

5

5 A headboard with all the works: lighting and ample storage keeps this space neat. **6** Among the beams of this Victorian space, a low-level bed looks inviting. While the massive wooden beams could dominate this space, the fact that everything else is close to the floor brings your eye level down.

7 In the roof space of an old factory, a skylight brings in ample light. Built-in storage runs the length of the old beam under the eaves, while vintage drawer units by Florence Knoll house the small items.
8 In this tiny room the bed runs along one wall. The custom-built headboard accentuates the height of the room, as do the wall applique, set at a low level. The pillows and shag-pile rug add a coziness.
9 Here a collection of mid-20th-century furniture is grouped. The headboard, night stand, and drawer unit are all by George Nelson, as is the book shelf at the end of the bed. Designed in the early 1950s, it couldn't be any more modern. The chair in the foreground is by Ralph Rapson.

may modernize, the floor in the sleeping zone is a very important element. To step out onto a rough sisal floor or a soft sheepskin rug is all a matter of personal taste, but in this zone it is often overlooked.

If you are in the situation of reorganizing your space, think about everything you will need in this space. Using a wall of storage to divide the sleeping zone from the bathing zone will help solve storage problems in both zones. A sleeping platform could be included by raising or lowering floors. If you are able to extend into your attic, this is perhaps a good time to create the ultimate sleeping zone. Being able to use one floor of your home for this personal zone would create a luxurious area that includes a bath area and all the requirements you associate with this zone.

Whatever way you modernize your home, in this zone the constraints are much the same: to create a suitable area to sleep, to be a retreat from the outside world, and to make a private space within a sociable home.

9

BATHING

I In this bathroom designed by Anne Fougeron for her own use, she has created a place she wants to spend time in. With its glazed roof and wall, one is enclosed with the elements outside and above.

In the home the bathing zone could be a shower room, a family bathroom, a wet room, or a steam room, or include a jacuzzi. What you ideally want and what you end up with does, of course, depend on the space you have and your budget. It also depends on your plumbing facilities and your overall needs in your home.

In the bathroom you need privacy, so the position within the total scheme is very important. If you are just decorating or updating for a more modern feel, you may have to put up with the existing placement of this room. But if the bathroom is on a level away from the bedrooms, or uses too large a space, relocating is understandable. You can now quite easily relocate wet areas using a macerator toilet that uses a very small pipe for waste. But to do this work is unnecessary unless your existing facilities are very inconvenient.

2 A suspended mirror screens off the bathtub. It houses a concealed light. It also opens up for ample storage. The sink unit is made from stainless steel. Overall the walls are covered with sheet glass. **3** The towel storage is faced with very thin laminated marble that creates a translucent panel that is illuminated at night. More stainless steel has been used for low-level storage, while the floor is poured terrazzo.

Today there is so much to choose from in the bathroom market. There are space-saving systems, built-in, freestanding and wall-hung bathtubs, basins, and toilets that make the choices endless. I suppose one of the worst scenarios is that you inherit an oddly colored bathroom from previous owner. This would have to go, unless you are brave and can work out a scheme to include and update it. If you are lucky enough to have white fixtures, these can usually be cleaned up, whatever condition they are in, and new wall finishes, faucets and accessories can easily and quickly make this zone sparkle. These days, with so many materials for wall finishes in wet areas available, there is no excuse for using the obvious. Even though most people's preference is to have white fixtures with white tiles, a bit of imagination can create a personalized space.

If you are reorganizing or extending your home, the placing of the bathroom or rooms can be up to you. To create a convenient wet area in a small unit can

4 In a bathroom in a house in London by Curtis Wood architects, different levels—the bench, the sink unit, and the storage unit—give ample space for a family bathroom. The wood used in this wet room needs regular maintenance to protect its surface. **5** A special paint finish has been used on the plaster walls that makes them totally waterproof. The floor has a hidden gully to take away the shower water. Above is a glass panel that brings light into this top-floor room. **6** In the second bathroom, a freestanding bathtub has been installed. While a shower is ideal for everyday use, sometimes we need a soak.

save space. But you may also need a separate bathroom to avoid having guests traipse through your bedroom. If you have a family, you will obviously need a family bathroom as well. If you have two wet areas, they can be back to back or side by side to save on plumbing costs. They may look better if they are treated the same, with the same fixtures and tiles or wall finishes. You could always use different towels and accessories to make them more suitable for the family.

The bathing zone should also carry on some sort of theme from the main living space. Use the same type or size of tile or faucets as in the kitchen, for example. Use the same lighting as in the hall, or the main bedroom when installing a connecting bathroom, and perhaps include similar built-in storage in the bedroom or dressing area to unite the spaces. The idea is to create a natural

6

7   8

flow from one space to another. Using similar color flooring, from wood to tile, also visually unites the space.

An extension could create an amazing bathing zone or wet space. You can build a box specifically as a bathroom perhaps with no windows but with skylights, with sunken areas for shower waste, or with central drainage, so the entire space becomes a shower room. You could also create a room with a view, so that you can bathe without being overlooked or so you look out over the bay or across the rooftops.

If you have children, you need to take them into consideration when working out the layout, scale of fixtures and the type of faucets that

need to be easy to use and easy to maintain. Bathrooms are an area of the home that are, no matter how well fitted out and decorated, often redone when a new owner comes along. The tiles are either the wrong color, there is a tub and you want a shower or vice versa, or in some cases it is in the wrong place. So when working on this zone, plan it with yourself in mind and make sure it works for you.

The bathing zone is an important part of today's home. It has to be private, yet work for your family and guests. It has to function well, with good water pressure for showers, and you need space to store all those bathroom and beautifying products.

7 This bathroom has totally opaque sliding windows, so there is no need for blinds or other window treatments. The mosaic wall adds great color to the space, and its compactness shows good planning. **8** Here a tiny bathroom has been squeezed between two bedrooms. The shape of the strip window is exaggerated by the use of the strip storage unit and shelf below. All the wall surfaces are painted. The toilet is wall mounted to increase the visual floor space. **9** The glass wall in this shower room has been sandblasted on the lower two thirds for privacy, so you can shower but still look out at the view without being noticed.

COOKING AND EATING

The kitchen is the most frequently updated and modernized zone of the house. One person's ideal is not always shared by the next.

In the home, the kitchen and eating areas are now often included within the main hub of the house. An open-plan space is usually worked around a central cooking and eating area, making this area a dividing space between the communal and private of the home. The simplest way to modernize or update this zone is to work with what you have. If you have good cabinets, update the doors, refresh the backsplash, and put in new faucets. All of this can be done quickly and easily and would cause little disruption. You can even update the appliances to a new finish or get integral versions that go behind closed doors. If you have culinary ambitions, perhaps you could rearrange the kitchen to cater for them. You can get industrial-looking stoves for the domestic market that suit these needs. This is maybe too extreme for a regular home kitchen, but it is an option if you intend to give this area heavy-duty use.

A cooking and eating zone should include ample work spaces for food preparation and for serving. It should have adequate appliances for your needs, the right size oven, one or maybe two dishwashers (one for clean dishes, one for dirty), and ample storage for cooking utensils and for food, both of the refrigerated and pantry type.

1 A well-stacked industrial shelf is the sign of a good cook, with a pot for every method of cooking.
2 In this Amsterdam home an industrial kitchen has been installed. An open island incorporates storage and an eating area and also houses a set-in sink. A wall of locker-type cabinets adds to the storage elements while racks around the stove keep everything at hand.

3 This open-plan kitchen has a bank of cupboards that protrude from the wall to form a peninsula unit. The simple white-painted units have a douglas fir countertop that relates to other wood elements found in the house. This kitchen leads to the central hallway. **4** The countertop has an in-set sink with a cooktop opposite. The wall cabinets hover just above the three-sided unit.

6

5 An overhang has been allowed to create an area for entertaining and somewhere to sit for a well-earned break. **6** In this kitchen designed for an active cook, an L-shaped island has been installed to give ample and necessary work surfaces. The reconstituted countertop offers a seamless and hygienic area for food preparation.

It is better if you can locate washing and drying services in their own dedicated area, but in this day and age, space can be at a premium so if these functions have to be included in the kitchen, there are clever decorative ways of integrating them within the cooking and eating zone.

To incorporate dining into this cooking space can mean that you need a breakfast bar area or a nearby dining table. You also will need storage for serving dishes and tableware, glasses and linens. The open-plan layout is good for the easy living ways of today. But of course if you are cooking in the

7 This grown-up kitchen, with its wood-faced units and exposed brick walls, makes the most of the bare materials of the old building. **8** The dining area in the same loft is in the central area of the space. The atrium nearby brings in light from above and adds another eating area, this time outside. The glass table top does allow it to disappear somewhat.

middle of the living area you will need a very powerful extractor hood to remove all cooking fumes at the source.

Today, the more formal dining room has been taken over by a more relaxed arrangement. Even though the set-up of a dining table and chairs does look good within the home, to give an entire room over to this is not the way forward. A more casual approach to eating has come into our lives, and in some cases this table arrangement becomes a work table during the day and a dining table at night.

If you are reorganizing your space, this cooking/eating zone can be located anywhere. If you position it toward the center of the space, it means you can access it from all areas and it truly becomes the heart of your home. But it is also good to have this relaxing or eating zone opening directly onto an outdoor space, which will increase the usable space considerably during the summer months.

The layout of the cooking area does depend on the space available. I always like to create some kind of island for the main appliances and sink areas. I find that when you have guests you can get a great group of friends around this space, and it also makes an ideal location for serving food and drinks. Adding a couple of stools at the correct height will also make this island a casual eating area.

Then it comes down to the style and look of the zone. For me, a plain simple kitchen with white doors and counters with concealed appliances and an integral sink gives a timeless classic modern kitchen area that will sit happily within any open-plan environment, whatever the style ethos. If you want to personalize this basic palette, or pick up on themes used elsewhere in the adjoining spaces, you can add color and texture with the careful choice of appliances and the occasional designer accessory.

9

Lighting in these zones is also very important. A well-illuminated cooking area is obviously needed, with concealed lighting above the cooktop. An extractor will probably include lighting as well. A well-lit kitchen area is usually desirable, but it is great to have a dimmer to give ambient light at night after the cooking is done—you do not want to floodlight the dirty dishes. In the dining area a great vintage pendant light can make the space, and well-directed concealed spotlights can also create a great atmosphere.

The furniture selected for this area is a matter of personal taste and style. Going for completely modern may mean going to the chain stores for great reasonably priced design, trawling through thrift stores or antique centers for good vintage examples, or going straight to the top designers for good new iconic pieces, perhaps specially commissioned.

The cooking-eating zone is the hub of the house. It needs to work, to be visually exciting, and it needs to be modern.

10

9 In this high-ceilinged kitchen, the glass doors of the wall cabinets mean everything is on display. Every cupboard needs to be well thought out and contain interesting items, while the glass doors need to be spotless. If you are not this well organised, perhaps too much display might not be a good thing! **10** The central island always works well, especially if enough space is allowed around it. In this kitchen there is ample space for a dining arrangement of tables, long benches and chairs between the island and the staircase. **11** In this wide bay window, built-in seating creates a space-conscious dining area.

WORKING

At home we are working more and more. From our home office we are designing magazine layouts, making one-off handbags and writing books. The homework space has firmly made its mark on our day-to-day living. It is now such a part of 21st-century living that it has to be included, these days, in all schemes we are working on.

Today working from home is commonplace. Whether you use the space under the stairs, an alcove in the main living area, or a recess in the hall, it is a value-added asset. And whichever way you modernize your home, the workspace needs to be included. To have a spare room that you can turn into a home office is a real luxury. To have a small area of your home that you can make into an work space is still a bonus. There is no real need to have this space hidden, but sliding partitions around it that enable you to hide the work chaos is somewhat of a relief when you have guests.

These days there are many ways to create the home office zone. Using vintage storage, you may be able to use part of a wall unit for your desk and computer components. At home I have a small mezzanine floor that I use as a library and workspace. Books in the home are usually the biggest and heaviest bulk of stuff we have and display, so to build or install an appropriate place for books and magazines is very important. To include this in a working zone is ideal. It is surprising how much stuff we do hoard.

As time goes by, I'm sure we will be working from home more and more, as this culture becomes more deeply entrenched. The space taken up for the home office will at least double, and the amount of time spent in the home will surely increase. The end of the 20th century saw this slowly happening; in this century more and more people are looking for quality of life, and working from home just adds to this.

1 In the corridor of this Florida house endless sliding doors closet a hideaway office. It's there when you need it, invisible when you don't.
2 This office is divided from the main living space by a wall of storage on one side and a sliding partition on the other, which allows it to be closed off when necessary.

3 Karim Rashid's New York homework area is basically a laptop on a glass table. This is as simple as it can be, but he includes other items such as good lighting and a sound system.

SUPPLIERS

FLOORING

Ariostea High Tech
High-quality marble and stone flooring.
www.ariostea-hightech.com

BuildDirect
Wholesale prices for laminate, bamboo, wood, and tile flooring.
(877) 631-2845
www.builddirect.com

Eco-Friendly Flooring
Wholesale supplier of environmentally friendly flooring products, including bamboo, cork, recycled glass tile, stone, and reclaimed and sustainable woods.
(866) 250-3273
www.ecofriendlyflooring.com

Gavin Historical Bricks
Reclaimed antique paving and building material.
(319) 354-5251
www.historicalbricks.com

Hosking Hardwood Flooring
Sales, installation, and renovation of hardwood floors.
(508) 643-0810
www.hoskinghardwood.com

Junkers
Solid hardwood floors in many woods and plank widths.
(714) 777-6430
www.junckershardwood.com

Pergo
One of the largest producers of laminate flooring in the world.
(800) 33-PERGO
www.pergo.com

The StoneYard
Natural stone and stone products.
800-231-2200
www.stoneyard.com

Wood Flooring International
Wood flooring sourced from around the globe; exotic (to North America) species and selected straight grained/quartered offerings in the common North American woods.
(856) 764-2501
www.wflooring.com

TERRAZZO:
The National Terrazzo & Mosaic Association, Inc.
Complete listing of terrazzo and mosaic suppliers.
www.ntma.com

Portal Terrazzo
On-line previews of terrazzo options.
www.portalterrazzo.com

CARPET:
Stainmaster
Carpet that specializes in resisting stains, ease of cleaning, standing up to traffic, and eliminating static shock.
(800) 438-7668
www.stainmaster.com

RUGS:
Alimadia Gallery
Traditional and contemporary hand-knotted rugs from Armenia, Turkey, Persia (Iran), Afghanistan, Tibet & Nepal, India, and Central Asia.
(507) 645-1651
www.alimadia.com

Chanadet.com
Decorative antiques and vintage artwork, objects, collectible books and textiles.
(212) 873-5983
www.Chanadet.com

STAIRCASE

Arcways
Custom curved and spiral staircases.
(800) 558-5096
www.arcways.com

Arke Stairs
Spiral and modular stair kits.
(888) 782-4758
www.arkestairs.com

The Iron Shop
Metal and wood spiral stair kits.
(800) 523-7427
www.theironshop.com

Stair Supplies
Supplies for any type of staircase, from treads to newels.
(866) 226-6536
www.stairsupplies.com

DOORS AND WINDOWS

Copper Moon Woodworks
Hand-made shutters that act as sophisticated, highly visible exterior design elements.
(610) 434-8740
www.coppermoonwoodworks.com

Doors By Decora
Decorative hardwood doors, standard or custom-bullt, with or without leaded and beveled glass.
(800) 359-7557
www.doorsbydecora.com

GlassWerxx
Stained-glass window shutters, oak interior shutters, and framed stained-glass panels.
(315) 493-0091
www.glasswerxx.com

Simpson Door Company
High-quality wood doors.
(800) 952-4057
www.simpsondoor.com

Velux
Specialists in roof windows and skylights, with a wide range of accessories including blinds and window-control options.
(800) 88-VELUX
www.velux.com

Windsor Windows and Doors
Quality windows and patio doors.

(800) 218-6186
www.windsorwindows.com

FIREPLACES

Crea France
Antique to contemporary fireplaces and mantels, restoration and installation.
(212) 213-1069
www.creafrance.com

Lennox Hearth Products
Fireplaces, freestanding stoves, fireplace inserts, and gas log sets.
www.lennoxhearthproducts.com

LIGHTING

Lutron
Products to control both natural daylight and electrical lighting.
(888) LUTRON1
www.lutron.com

Rejuvenation
Recreations of traditional American lighting and hardware.
(888) 401-1900
www.rejuvenation.com

Restoration Hardware
High quality textiles, furniture, lighting, bathware, hardware and amusements.
(800) 816-0901
www.restorationhardware.com

Shades of Light
Specializes in unique lamps and lighting with many original designs.
(800) 262-6612
www.shadesoflight.com

Verilux
Lighting solutions that simulate natural light, including task, ambient and direct lighting applications.
(800) 786-6850
www.healthylight.com

DISPLAY, SHELVING, STORAGE

California Closets
Customized storage solutions designed for your space, available nationwide.
(888) 336-9709
www.calclosets.com

The Container Store
Helping people streamline and simplify their lives by offering an exceptional mix of storage and organization products.
(888) CONTAIN
www.containerstore.com

Hold Everything
The "Masters of Stylish Organization" feature products to make organization easy.
(800) 421-2264
www.holdeverything.com

Ikea
Large selection of affordable modular and freestanding wardrobes and storage systems.
(800) 434-4532
www.ikea.com

WALLS

Benjamin Moore
A wide selection of paint colors and finishes, available nationwide.
(800) 344-0400
www.benjaminmoore.com

Carter and Company
Reproductions of significant historic wallpapers, from the early-19th through the mid-20th century, both European and American.
(707) 554-2682
www.carterandco.com

Crossville
Porcelain stone tile, ranging from traditional styles to a line made with metals such as bronze, copper, and brass.
(931) 484-2110
www.crossville-ceramics.com

Devine Color
A vibrant paint line that focuses on the relationships between color and light.
(866) 926-5677
www.devinecolor.com

Farrow & Ball
Muted and historical paint colors in more than 100 shades and reproduction wallpaper.
(845) 369-4912
www.farrow-ball.com

Sherwin-Williams
Nationwide stores offer over 1000 color choices and a large selection of wallpaper.
www.sherwinwilliams.com

FURNITURE

NEW:
Ethan Allen
Home furnishing solutions for every room in your home.
888-EAHELP1
www.ethenallen.com

Ikea
New designs and good basics.
(800) 434-4532
www.ikea.com

Pottery Barn
Furniture, bedding and bath, rugs, lighting, and window treatments for the home.
(888) 779-5176
www.potterybarn.com

Thomasville
Furniture, upholstery, cabinetry, and lighting for your home.
(800)225-0265
www.thomasville.com

REPRODUCED & ORIGINAL CLASSICS:
American Accents
Mission, Shaker and Country reproductions.
(336) 885-7412
www.americanaccentsfurniture.com

Designer Antiques
Antiques and custom-designed wood furniture.
(800) 261-0283
www.designerantiques.com

Fritz Hansen
Timeless, functional, minimalistic and innovative quality furniture, including reproductions of Arne Jacobsen and Poul Kjareholm furniture.
(646) 495-6183
www.fritzhansen.com

HermanMiller
Furniture for living and work environments, including Eames office reproductions.
(888) 443-4357
www.hermanmiller.com

Knoll
Furniture design, including reissues of work by Florence Knoll, Harry Bertoia, and Eero Saarinen.
877-61KNOLL
www.knoll.com

L.W. Crossan
18th century reproductions, including William and Mary, Queen Anne, Chippendale, Hepplewhite, and Sheraton styles.
(610) 942-3880
www.lwcrossan.com

Modernica
Reissues of many Eames and Nelson designs.
www.modernica.net

1stDibs.com
International listing of decorative arts dealers worldwide.
www.1stDibs.com

KITCHENS

Bisazza
Hard-wearing and sleek surfaces for bathroom worktops, walls, and sinks, in many finishes and colors.
(305) 597-4099
www.bisazza.com

Corian
Long-wearing and sleek surfaces for countertops, walls, and sinks in many finishes and colors.
(800) 4-CORIAN
www.corian.com

Formica
Designers and manufacturers of all types of surfacing materials, including laminate, metal, wood, stone, solid surfacing, and sinks.
800-FORMICA
www.formica.com

SileStone
Natural quartz surfaces that are heat-, scratch-, and scorch-resistant.
www.silestone.com

Vermont Soapstone
Custom manufacturer of soapstone countertops, sinks, and fireplaces.
(802) 263-5404
www.vermontsoapstone.com

BATHROOMS

American Standard
The world's largest producer of bathroom and kitchen fixtures and fittings.
(800) 442-1902
www.americanstandard-us.com

Ceco
Porcelain enameled cast-iron plumbing fixtures.
(323) 588-8108
www.cecosinks.com

Clawfoot Supply
Large selection of clawfoot tubs, as well as unique and hard to find items such as pedestal sinks, console sinks, shower curtain rods, pull chain toilets, and curved shower rods.
(877) 682-4192
www.clawfootsupply.com

Kohler
Luxury bath products designed for gracious living.
(800) 4-KOHLER
www.kohler.com

Mansfield Plumbing Products
Attractively designed, high performance plumbing fixtures and fittings.
(419) 938-5211
www.mansfieldplumbing.com

Sculptured Homes
Manufacturer of the WetSpa, a luxury, frameless, glass steam shower enclosure that incorporates stereo sound and fiber-optic lighting.
(877) WET-SPAS
www.sculpturedhomes.com

Waterworks
Luxury bath supplies from fixtures, fittings, and furnishings to tile, towels, and apothecary.
www.waterworks.com

ARCHITECTS

Ulla Koskinen
Untamonte 4a 00610
Helsinki
Finland
003 58986835450
koskinen.rantanen@sci.fi

MIKKO PUOTILA RESIDENCE, HELSINKI:
19 right, 93 right, 98 bottom left, 120 left,
122, 144, 145.
RITVA PUOTILA RESIDENCE, HELSINKI:
7 left, 52.
NEILAMA RESIDENCE, HELSINKI:
20, 32, 33, 38, 118 right, 130 bottom left,
166 right.

Kari Lappalainen
003 5896801828
kari.lappalainen@pp7.inet.fi

ILLKA APARTMENT, HELSINKI:
98 bottom middle, 124 right, 140, 142
bottom right, 153 right, 153 left, 155 bottom
left, 163.

Gene Leedy
555 Avenue G NW
Winter Haven
FL 33880
001 863 293 7173
www.geneleedyarchitect.com

JUNIOR WHITTINGHILL RESIDENCE, WINTER HAVEN,
FLORIDA:
10, 102 left, 176 bottom right, 177.
ROBERT KAISER RESIDENCE, WINTER HAVEN,
FLORIDA:
50, 51, 118 middle, 123 right, 146, 147,
157.

Anne Fougeron
Fougeron Architecture
720 York St Suite 107
San Francisco
CA 94110
001 415 641 5744
anne@fougeron.com
www.fougeron.com

FOUGERON RESIDENCE, SAN FRANCISCO:
6, 39 top left, 44, 45, 46, 47, 96 bottom
right, 114 left, 115 bottom left, 121, 130
right, 134 right, 138 left, 148, 149, 168, 169.

Curtis Wood Architects
Andrew Wood/Jason Curtis
The Shopfront
84 Haberdasher Street
London N1 6EJ
020 7684 1400
www.curtiswoodarchitects.com
info@curtiswoodarchitects.com

STARK RESIDENCE, PUTNEY, LONDON:
3, 14, 54, 55, 56, 57, 73, 97 right, 103 left,
109 top right, 142 top right, 170, 171.

Andrew Weaving
68 Marylebone High Street
London W1
modern@centuryd.com
www.centuryd.com
07808 727615

WEAVING/THOMASSON RESIDENCE, LONDON:
30, 31, 70, 71, 99, 102 right, 109 bottom
right, 155 top left,166 left, 172 right.
WEAVING/THOMASSON RESIDENCE, ESSEX:
21, 26, 82, 83, 84, 85, 96 top right, 112
right middle, 128, 167, 155 bottom right,
167, 176 top left, 176 bottom left.

Marc Prosman Architecten Bv
Overtoom 197
1054 Ht Amsterdam
31(0)20489 2099
Fax 31(0)2048 3658
architecten@prosman.ni
www.prosman.ni

VAN BREESTRAAT, AMSTERDAM:
1, 5, 7 middle, 60, 61, 62, 63, 95, 126, 158,
182 right.
ALBERDINGH THIJMSTRAAT, AMSTERDAM:
58, 59, 97 left.
SURINAMPLEIN, AMSTERDAM:
93, 94, 112, 115 top right,155 top right.

Next Architects
Weesperzijde 93
1091 Ek Amsterdam
31(0)20463 0463
Fax 31(0)203624745
info@nextarchitects.com
www.nextarchitects.com

DE STAD, AMSTERDAM:
123 bottom left, 152, 174, 175.
PRINCENGRACHT, AMSTERDAM:
153 middle.

Don Chapell (Deceased)
Guy Peterson Faia
Guy Peterson/Ofa Inc
1234 First Street
Sarasota
FL 34236
001 941 952 1111
www.guypeterson.com
guypeterson@guypeterson.com

FISHMAN RESIDENCE, SARASOTA, FLORIDA:
8, 16, 19 left, 19 middle, 74, 75, 76, 77, 78,
79, 96 left, 104 right, 136, 139, 142 left,
164 left.
Interior by Wilson Stiles, Sarasota, Florida.

Ogawa Depardon Architects
137 Varick Street No. 404
New York
NY 10013

212 627 7390
Fax 212 627 9681
www.oda-ny.com

HOUSE IN BROOKLYN, NEW YORK:
13, 64, 65, 88, 89, 90, 91, 92, 129, 133, 135.
Interior By Michael Formica.
HILPERT RESIDENCE, BELVEDERE, SAN FRANCISCO:
12, 100, 101, 111, 114 right top, 114 bottom
right, 116, 117, 125, 131, 143, 162, 173.
NIPON APARTMENT, NEW YORK:
18, 86, 87, 124 left.

Downtown Group
236 West 27th Street No. 701
New York
NY 10001
212 675 9506
info@downtowngroup.com
www.downtowngroup.com

KANETELL RESIDENCE, TRIBECA, NEW YORK:
2, 11, 80, 81, 106, 107, 118 bottom left,
141 left, 159 right, 178, 179.

1100architect.com
435 Hudson Street 8th Floor
New York
NY 10014
212 645 1011
Fax 212 645 4670
info@1100architect.com

SHANE/COOPER RESIDENCE, NEW YORK:
23 right, 28, 29, 53, 66, 67, 68, 69, 104
bottom left, 105, 134 left.
AND ALSO:
109 bottom left

Karim Rashid
357 West 17th Street
New York
NY 10011
001 212 929 8657
info@karimrashid.com
www.karimrashid.com

OWN APARTMENT, NEW YORK:
22, 23 left, 24, 25, 161, 183.

Seibert Architects
Sam Holladay
352 Central Avenue
Sarasota
FL 34236
001 941 366 9161
www.seibertarchitects.com
Fax 941 365 0902

SAM HOLLADAY RESIDENCE, SARASOTA, FLORIDA:
9, 103 bottom right, 114 top left, 115 top
left, 119 top left, 172 top left.

Others

GEORGE RESIDENCE:
Extension remodeling by Michael George
01225 850561
72, 93 middle, 112 bottom right, 120 right.

MANNISTO/POYHONEN APT, HELSINKI:
Interior by Tuula Poyhonen
tuula.poyhonen@fonet.fi
15, 34, 35, 89 top right,129 top left, 138 right.

LINCOLN/ORUM RESIDENCE, SUFFOLK:
Interior by Angi Lincoln
07957 621796
7 right, 36, 39 bottom left, 48, 49, 104 top
left, 109 top left, 113 left, 129 right, 132,
137 left, 165 right.

SPRINGMAN WESTOVER RESIDENCE, LONDON:
130 top left, 151

HARDING RESIDENCE:
Architecture by Audrey Matlock Architects
180 left, 181 right.

MACKERETH/VOGEL RESIDENCE, LONDON:
Designed by Wells Mackereth
98 top left, 127, 180, 181 middle.

COLLINS RESIDENCE, LONDON: *150.*

RIOS RESIDENCE, LOS ANGELES: *110.*

BOUCQUIAU RESIDENCE, BELGIUM:
Designed by Marina Frisenna
108 right.

INDEX

ACKNOWLEDGMENTS

First of all I would like to thank Jacqui Small for having faith in this book idea and getting it off the ground during a not-so-easy time. Thanks to Kate John and Ashley Western for getting it all together and making a great job with the selection of locations and my garbled text.

Of course it's the friends, clients, architects, and designers who have made it so much easier and fulfilling. Thanks to all of you. Being able to enter other people's homes, shuffle their possessions around, and take over their lives for a day has made this book the way it is. To Angi and John and Antony and Eric—thanks for your patience and time. To Rob Bevan, I look forward to your new home, and thanks to Barbara and Mike George for letting us in, even though the job was not completed.

In the US, without the help, time, and enthusiasm of Gilles Depardon, many of the pages in this book would be blank; also thank you Anne Fougeron for letting us into your home and for creating such a great photogenic house. In New York, thanks to Jo Shane and John Cooper for opening your doors to us. Many thanks to Karim Rashid for letting us free in your space. And of course in Florida, thanks initially to Martie Lieberman for putting up with my long-winded emails of questions and information! Without Martie, the great places in Sarasota would not have been so accessible. Thanks to Guy Peterson, Sam Holladay, and Junior Whittinghill and Robert Kaiser of Winter Haven. And of course, thanks so much to Les and Lois Fishman for your hospitality and time. We will be back!

In Finland, Ulla Koskinen ran around town showing us the great interiors featured here, promoting not only her own work but also that of the local competition. My thanks Ulla. In Amsterdam, Marc Prosman let us into at least three of his projects, all different, and all included here. Thank you Marc. And thanks to Next Architects for organizing access during our very short stay there.

Back home, thanks to Ian again for putting up with the paperwork, for being there before and after my trips, and for just telling me to get on with it when necessary.

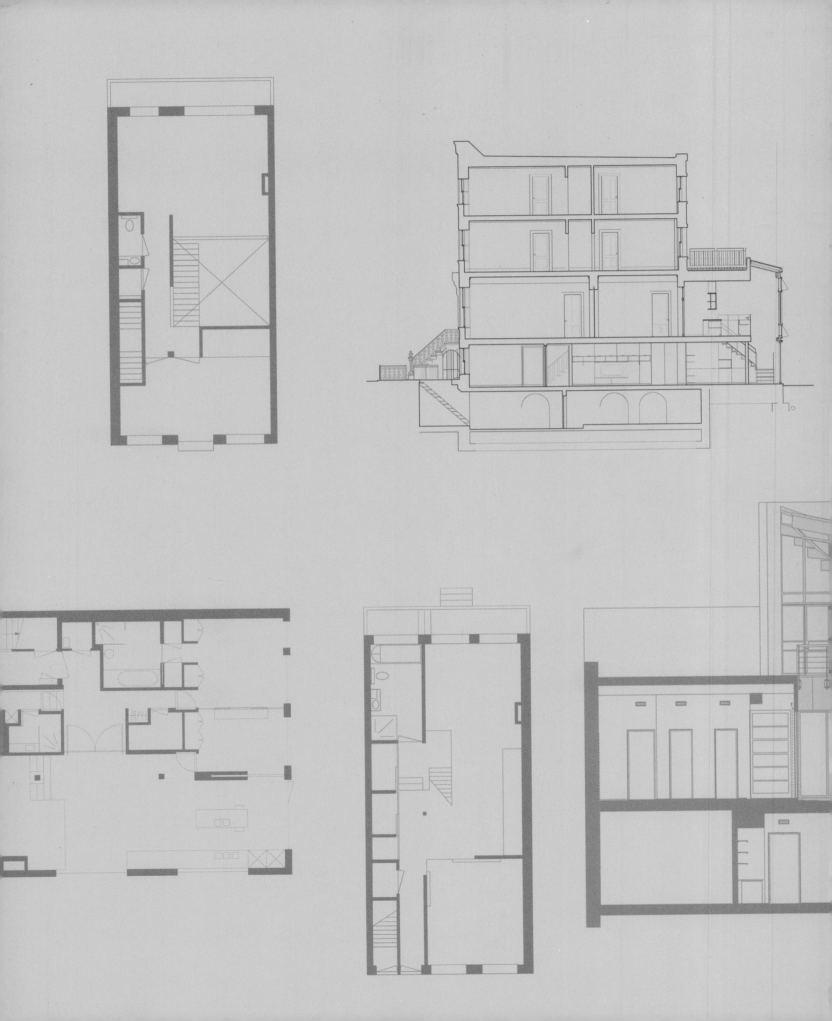